HISTORICAL CONNECTIONS IN MATHEMATICS VOLUME II

Resources for Using History of Mathematics in the Classroom

Wilbert Reimer
Fresno Pacific University

Luetta Reimer
Fresno Pacific University

Brenda Wood, Illustrator
Roxanne Williams, Desktop Publisher

AIMS Education Foundation
Fresno, California

This book contains materials developed by the AIMS Education Foundation. **AIMS** (**A**ctivities **I**ntegrating **M**athematics and **S**cience) began in 1981 with a grant from the National Science Foundation. The non-profit AIMS Education Foundation publishes hands-on instructional materials that build conceptual understanding. The foundation also sponsors a national program of professional development through which educators may gain expertise in teaching math and science.

Copyright © 1993, 2005 by the AIMS Education Foundation

All rights reserved. No part of this book or associated digital media may be reproduced or transmitted in any form or by any means—including photocopying, taping, or information storage/retrieval systems—except as noted below.

- A person or school purchasing this AIMS publication is hereby granted permission to make up to 200 copies of any portion of it (or the files on the accompanying disc), provided these copies will be used for educational purposes and only at one school site. The files on the accompanying disc may not be altered by any means.

- Workshop or conference presenters may make one copy of any portion of a purchased activity for each participant, with a limit of five activities per workshop or conference session.

- All copies must bear the AIMS Education Foundation copyright information.

AIMS users may purchase unlimited duplication rights for making more than 200 copies, for use at more than one school site, or for use on the Internet. Contact us or visit the AIMS website for complete details.

AIMS Education Foundation
P.O. Box 8120, Fresno, CA 93747-8120 • 888.733.2467 • aimsedu.org

ISBN 978-1-60519-038-9

Printed in the United States of America

Table of Contents

INTRODUCTION

CHAPTER 1
Portrait of Thales (c. 636 - c. 546 B.C.) 1
Thales: The Problem Solver 2-3
Activities:
    Smart Shadows 4
    Stars Around The Moon 5
    Puzzling Pyramids 6
    Dominoes on a Checkerboard 7
    Who's Who on the Baseball Team? 8

CHAPTER 2
Portrait of Euclid (c. 330 - c. 275 B.C.) 9
Euclid: The Father of Geometry 10-11
Activities:
    Euclid's Algorithm 12
    The Greatest Common Divisor Meets The Least Common Multiple 13
    Perfect Numbers 14
    Daffy Definitions from Geometry 15
    White-Faced Cubes 16

CHAPTER 3
Portrait of Heron (c. 75) 17
Heron: The Ingenious Engineer 18-19
Activities:
    Heron's Square Root Method 20
    Heron's Formula 21
    The Chocolate Cake Challenge 22
    Space for Ace 23
    Diagonal Challenge 24

CHAPTER 4
Portrait of Hypatia (c. 370 - c. 415) 25
Hypatia: Model of Excellence 26-27
Activities:
    The Amazing Ellipse 28-29
    Changing a Quarter 30
    Paper Folding A Parabola 31
    A Stamp Stumper 32

CHAPTER 5
Portrait of Banneker (1731 - 1806) 33
Banneker: Self-Taught Genius 34-36
Activities:
    Two Puzzles from Benjamin Banneker's Collection 37
    A Clockmaker's Challenge 38
    A Collection of Puzzlers 39
    Make-A-Square Puzzle 40
    A Jumble Puzzle 41

CHAPTER 6
Portrait of Babbage (1792 - 1871) 42
Babbage: Making a Difference 43-44
Activities:
    Rings of Circles 45
    Pizza Party 46
    Tournament Time 47
    Region Revenge 48
    Checkerboard Challenge 49

CHAPTER 7
Portrait of Galois (1811 - 1832) 50
Galois: The Misfortunate Mathematician 51-53
Activities:
    Rotating Tires: A Mathematical Event! 54
    Clock Arithmetic 55
    Magic Flexagons 56-57
    The Erased Chance: A Skit to Read 58-59

CHAPTER 8
Portrait of Lovelace (1815 - 1852) 60
Lovelace: The First Computer Programmer 61-62
Activities:
    Calculator Fun 63
    Circle Circus 64
    Twenty-One Connect 65
    Leap Frog 66
    Round Robin Play 67

CHAPTER 9
Portrait of Kovalevsky (1850 - 1891) 68
Kovalevsky: Daring and Determined 69-71
Activities:
    The Snowflake Curve 72
    Predictable Patterns 73
    The Chessboard Covered With Wheat 74
    Paper Punching Patterns 75

CHAPTER 10
Portrait of Ramanujan (1887 - 1920) 76
Ramanujan: India's Mathematical Genius 77-78
Activities:
    Counting Partitions 79
    The Curious Cab 80
    Finding Pi 81
    Numbers: You Gotta Love 'Em 82
    Fun With Numbers 83

APPENDIX
The Nine Digit Puzzle 85
The Magic Card Game 86

Making Magic Cards 87
A Crossword Puzzle 88
Mathematicians: A Card Game 89-102

SUGGESTIONS AND SOLUTIONS 103-115
RESOURCES FOR LIBRARY AND CLASSROOM 116-118

INTRODUCTION AND SUGGESTIONS FOR TEACHERS

"One can invent mathematics without knowing much history. One can use mathematics without knowing much, if any, of its history. But one cannot have a mature appreciation of mathematics without a substantial knowledge of its history."
-- Abe Schenitzer

Our goal in this series is to provide a collection of resources to make it easy for teachers to integrate the history of mathematics into their teaching. While mathematics history textbooks abound, there are not many sources that combine concise biographical information with activities to use in the classroom. We hope that the problem-solving experiences, the portraits, and the anecdotal stories will facilitate a broad, natural linkage of human elements and mathematical concepts.

The value of using history in teaching mathematics is currently gaining emphasis throughout the world and in the United States. Providing a personal and cultural context for mathematics helps students sense the larger meaning and scope of their studies. When they learn how persons have discovered and developed mathematics, they begin to understand that posing and solving problems is a distinctly human activity.

Using history in the mathematics classroom is often a successful motivational tool. Especially when combined with manipulatives, illustrations, and relevant applications, historical elements have the power to make mathematics "come alive" as never before. By viewing mathematics from a historical perspective, students learn that the *process* of problem solving is as important as the *solution*.

This book can be used in many ways. The teacher may choose to read or share biographical information and anecdotes as an introduction to one or more of the activities in a particular section. Portraits may be posted or distributed, and puzzles or problems may be used independently. It may be most effective, however, to focus on one mathematician at a time. A wide range of activities may be incorporated into a unit on a specific mathematician, allowing the teacher to make cross-disciplinary connections with social studies, language arts, and science.

Mathematicians may be selected for emphasis according to the concepts being introduced in the mathematics curriculum, or they may be selected at random for enrichment. While some of the activities do not replicate the exact problems the mathematicians worked on, they represent the areas of interest of those mathematicians. Activities have been chosen to appeal to a wide range of interests and ability levels.

A unique feature of this volume is the complete set of cards for playing "Mathematicians," a card game designed to review the mathematicians featured in Volumes I and II of *Historical Connections*.

Complete solutions for all the activities and specific suggestions for use are included in the back of this book.

Wilbert Reimer
Luetta Reimer

SOME GENERAL SUGGESTIONS ON HOW TO INTEGRATE MATHEMATICS HISTORY INTO THE CLASSROOM

Through reading aloud:
Students respond very positively to simply hearing stories read aloud. Build a collection of brief and interesting stories to read in the classroom. *Mathematicians Are People, Too* was written with this purpose in mind, but there are other sources as well. Enlarge and photocopy the illustrations in *Historical Connections in Mathematics* onto transparencies, displaying them at appropriate times during the reading.

Through writing:
Many options for writing projects arise from mathematics history. Students may research the life of a particular mathematician and write a report. They might read a biography or a historical novel about a mathematician and write a book review. They may be asked to write an imaginary interview, a newspaper story, a screenplay, or a poem about an individual from mathematics history.

Other possibilities include writing about the origin of a particular concept or symbol, such as the = sign or π. Some might wish to write about how mathematics was understood in a particular period or a particular place.

Through skits or video productions:
Choose a mathematician for special focus. Read to the class a brief biographical sketch or a collection of anecdotes about the mathematician being studied. We suggest the information in *Mathematicians Are People, Too* or *Historical Connections in Mathematics Volumes I and II*. Older children may read and research on their own. Invite small groups of students to prepare a skit or, if equipment is available, a video to share with the class.

Through hands-on experiences:
Many of the activities in this book invite hands-on experiences. While some can simply be done using paper and pencil, students will become more involved and remember the concepts better if they can participate more fully in the process. Encourage them to build models and to conduct experiments. Reinforce the principles of the scientific method as a problem-solving tool.

Through the arts:
Challenge students to draw or paint a scene from the life of the mathematician being studied. Remind them of the importance of researching the architecture, clothing, furnishings, etc., of the time and place in which the mathematician lived.

Ask students to write a poetic ballad or lyrics to a song that incorporates the major life events/accomplishments of a particular mathematician.

Through visual aids:
Consider displaying time lines, posters, portraits, or quotations from famous mathematicians. Check out the options in films and videos. Build a collection of postage stamps from various countries that honor mathematicians and mathematical development. Bake a cake and celebrate a mathematician's birthday!

Thales
c. 636 - c. 546 B. C.

THALES
THE PROBLEM SOLVER

Biographical Facts:

Thales (THAY-leez) of Miletus lived from about 636 to 546 B.C. Very little is known about his appearance or personal life. He was a well-educated Greek who traveled widely throughout Egypt and Babylon. Some believe that Thales was a teacher of Pythagoras.

Contributions:

Thales, one of the "seven wise men" of antiquity, is the first known individual associated with mathematical discoveries. His contributions reflect a wide range of interests. He developed a reputation as statesman, counselor, engineer, businessman, philosopher, astronomer, and mathematician. Undergirding all of these areas was a compulsion to know things for himself. He practiced his approach—observing, studying the pattern, and predicting the natural path—in many situations. He was not content with the accepted notions of his time, which attributed almost everything to myth and superstition.

Quotations by Thales:

"Know thyself."

When asked how to be good, Thales responded: "By refraining from doing what we blame in others."

Anecdotes:

A Bumper Crop

One of Thales' friends was bemoaning his poverty. "It seems there's nothing a poor man can do to get ahead," he groaned. Thales tried to cheer him up. "If we really set our minds to it, we can be rich! Come visit me in six months and you'll see."

Thales began to look around his community. There were lots of poor people because the oil-producing olive trees, on which they depended for their livelihood, had not produced well for years. After investigating weather patterns, talking with old-timers, and checking out the trees, Thales predicted a bumper crop. He bought all the oil presses he could; people were willing to sell them at low prices. When the olives were harvested, the same persons had to rent presses from Thales to handle all the olives, and he made a fortune!

His point had been made: careful thinking and pattern observation may lead to riches. But Thales was more interested in the process than the result. He sold the presses back to the community, and went on to other pursuits.

The Frustrated Tour Guide

On one of Thales' trips to Egypt, he visited the ancient pyramids, including the great Cheops, which covers 12 acres. Curious as always, Thales asked a local guide how tall the pyramid was. No one knew! While they scrambled to locate the answer, Thales calmly surveyed the situation. Knowing his own height, he used the length of his shadow and the shadow cast by the pyramid to solve the problem.

Teaching a Donkey

Legend says that Thales once supported himself by operating a salt mine. A train of donkeys was used to transport the salt from the mine to the port where it could be processed and shipped out. To reach the port, the donkeys had to cross a shallow stream. One day a donkey lost its footing in the stream and fell into the water, dissolving the salt strapped to its back. Appreciating the lighter load, the donkey purposely "fell" almost every day, making it worthless to its owner.

Thales observed the pattern and found a solution: he had his workers strap sponges instead of salt onto the donkey's back. The next time the donkey hit the water, it scrambled out of the stream to find the load heavier than ever. After only a few days of wearing sponges, the donkey was cured! This story is recorded as one of Aesop's Fables.

The Beginnings of Geometry

Thales' teachers knew geometry of surfaces, and used what they knew to measure real estate. But Thales moved to the level of abstraction, creating for the first time in the history of the world a geometry of lines. He was the first to state the following theorems:
- A circle is bisected by its diameter.
- Angles at the base of an isosceles triangle are equal.
- When two straight lines cut each other, the vertically opposite angles are equal.
- The inscribed angle in a semicircle is a right angle.

The Dark Peace

Thales, always interested in astronomy, set out to disprove a common belief in his time that the sun was only about a foot in diameter. His "measurements" found it much larger than anyone before him had dared imagine, but still much smaller than it actually is.

The sun was at the center of many ancient myths, so it is easy to understand why an eclipse of the sun would create terror in the hearts of Thales' contemporaries. But he knew what caused eclipses, and predicted the day one would occur. The public scorned his predictions, and the Medes and Lydians, at war for six years, planned a spectacular battle for that day. When the sun disappeared at about the time Thales had announced, they were terrified, and the two armies signed a peace treaty.

Watch Your Step!

Plato tells a humorous story about Thales. One night the mathematician was out walking. Apparently, he was gazing at the stars and thinking about the size of the universe when he suddenly fell into a well. An old woman heard his cries and came to his rescue, but scolded him with contempt. "Here is a man who would study the stars," she sneered, "but cannot see what lies at his feet."

SMART SHADOWS

Thales impressed the early Egyptians by measuring the height of pyramids using their shadows.

Find the height of a flagpole, a tree, or some other object using Thales' technique.

Here is the procedure:
1. Place a stick vertically in the ground and measure its height.
2. Measure the length of the shadow cast by the stick.
3. Measure the length of the shadow cast by the flagpole or other object.
4. Multiply the stick's height by the length of the flagpole's shadow.
5. Divide your result by the length of the stick's shadow. This result is the height of the flagpole.

	Length of Shadow	Height
Stick		
Flagpole		
Tree		
Other		

STARS AROUND THE MOON

Thales was intrigued by patterns he found in science and nature. His friends liked to tell the story of the time he was looking so intently at the night sky that he fell into a ditch. Most of the time, his observation led to more positive results!

Use good observation in this activity to determine the number of "stars" around the "moons."

3	1	2	4	5	__6__
4	5	3	3	1	__8__
1	2	3	4	5	__6__
1	1	1	1	1	__0__
6	6	6	6	6	__0__
3	3	3	3	3	__10__
4	4	5	5	5	____
1	3	3	4	1	____
2	4	4	1	5	____
3	3	1	4	1	____
4	5	5	2	3	____

HISTORICAL CONNECTIONS VOL. II   5   © 2005 AIMS Education Foundation

PUZZLING PYRAMIDS

The great Egyptian pyramids fascinated Thales not only because of their magnitude, but also because of their shapes. He and other early Greek mathematicians saw whole numbers as geometric shapes. To explore this, Thales probably stacked a pile of round pebbles into a triangular pyramid.

Three popular pyramid puzzles can easily be made using practice golf balls and an electric glue gun. Construct the individual pieces and then solve the puzzles!

PUZZLE NO. 1

Use these five pieces to make a triangular pyramid.

PUZZLE NO. 2

Use these six pieces to build a triangular pyramid.

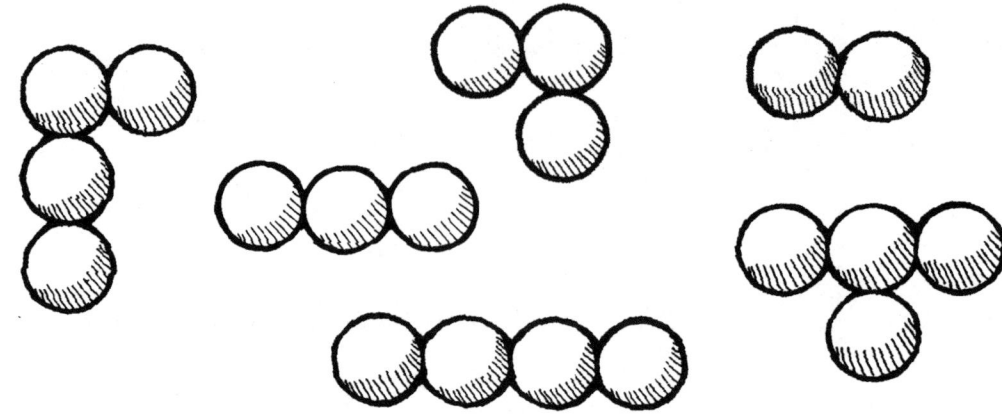

PUZZLE NO. 3

Build a 3 by 4 rectangular pyramid, using the same six pieces as in Puzzle 2.

DOMINOES ON A CHECKERBOARD

A domino is the same shape and size as two squares on a checkerboard, so a standard checkerboard can be completely covered with 32 dominoes. What happens if two opposite corners of a checkerboard are removed?

Thales became well-known for his deductive reasoning. Follow his example as you think logically about this problem.

Is it possible to cover the remaining checkerboard with dominoes? _____

Why or why not? _____

_____

_____

Would it be possible if only the two lower corners were removed? _____

Why or why not? _____

_____

_____

_____

HISTORICAL CONNECTIONS VOL. II　　　　　　7　　　　　　© 2005 AIMS Education Foundation

WHO'S WHO ON THE BASEBALL TEAM?
A Logical Thinking Activity

Use these clues to identify the position of each member of a baseball team. Check off positions players cannot hold in the grid below.

1. Jeff and the third baseman attend the same school.
2. Troy, Jeff, Luis, and the catcher were beaten in bowling by the second baseman.
3. Adam is a very close friend of the catcher.
4. The shortstop, the third baseman, and Luis like to go to the races.
5. Troy and Matt each won $5 from the pitcher.
6. Matt, Adam, Jeff, and the three outfielders go bicycling together.
7. Eric and the outfielders like skiing.
8. Troy, Paul, the center fielder, and the right fielder enjoy classical music.
9. Troy, Adam, and the shortstop are over 6 feet tall.
10. The first baseman, the shortstop, and Jeff live in the same city.
11. The left fielder, the third baseman, and Troy enjoy jogging.
12. The center fielder, the right fielder, and Luis wear glasses.
13. Adam and the second baseman often work out together.
14. The second baseman, the shortstop, and Eric love Italian food.
15. Paul and the shortstop grew up in the same town.
16. Lee and the center fielder are both left-handed.

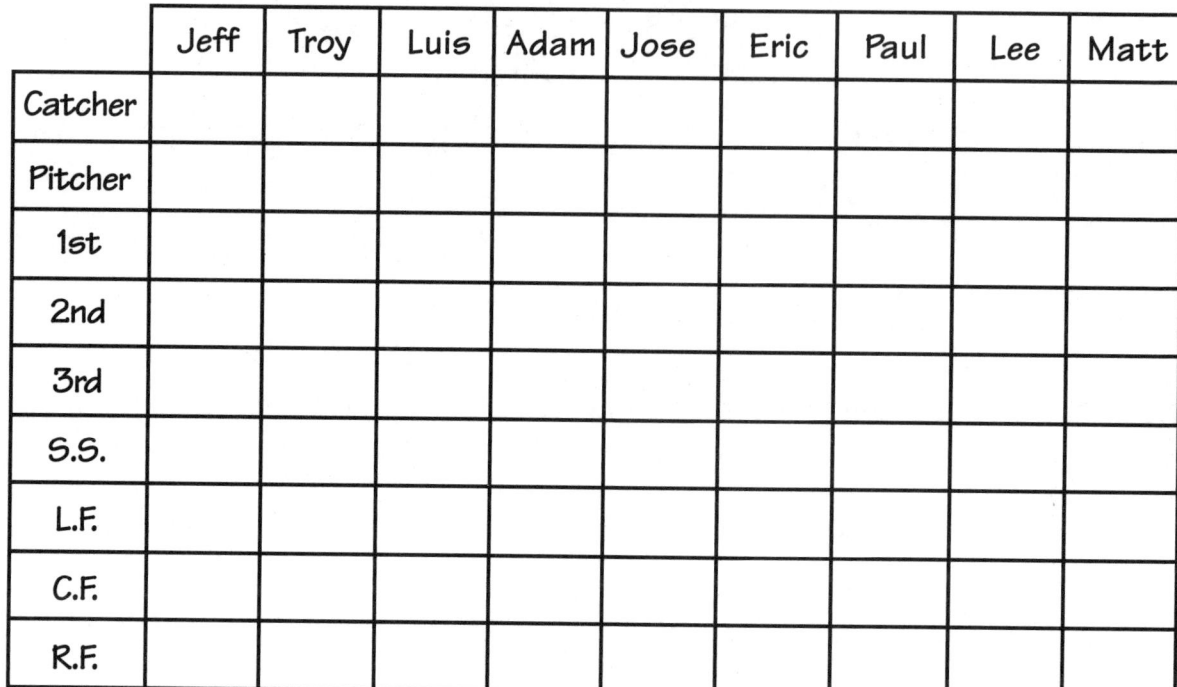

	Jeff	Troy	Luis	Adam	Jose	Eric	Paul	Lee	Matt
Catcher									
Pitcher									
1st									
2nd									
3rd									
S.S.									
L.F.									
C.F.									
R.F.									

HISTORICAL CONNECTIONS VOL. II © 2005 AIMS Education Foundation

Euclid
c. 330 - c. 275 B. C.

EUCLID
THE FATHER OF GEOMETRY

Biographical Facts:

Euclid (YOO-klid) was a Greek mathematician who lived from about 330 to 275 B.C. Very little is known about his personal background, except that he studied in Plato's academy in Athens before being invited to teach at the University of Alexandria.

Euclid was respected as a modest, friendly man, always kind and patient. His work shows that he was scrupulously fair, quick to acknowledge the contributions of others. He is also a model of accuracy, imagination, determination, and logical thinking.

Contributions:

Euclid is often called the "Father of Geometry." Although he was not a creative mathematical genius, he contributed greatly to mathematics through his expository skills. He wrote ten major works. The most important is *Elements*, dealing with algebra, geometry, and number theory. He organized the available mathematical knowledge of his day into a clear, orderly format. This book became the primary mathematical text for the next 2000 years.

Quotation by Euclid:
"There is no royal road to geometry."

Anecdotes:

Alexandria: Center of Learning

After the death of Alexander the Great in 323 B.C., his territories were divided. Egypt came under the rule of Ptolemy I, a benevolent man who embraced learning. He began immediately to make Alexandria into a cultural and educational center. He built the first university—remarkably like universities today—with lecture rooms, laboratories, gardens, museums, libraries, and living quarters. The library, at the heart of the complex, was said to contain over 700,000 papyrus rolls. The university opened its doors in about 300 B.C. and was the intellectual center of Greek life for nearly 1000 years.

Euclid was invited to head the department of mathematics at the university. There he enjoyed working with the finest scholars in the world.

The Second Greatest Book

Euclid's *Elements* was comprised of 13 sections. The original work filled 133 rolls of parchment.

Most of the postulates, theorems, and axioms in the book were not original with Euclid; the genius of the book is in its organization. Euclid was able to compile all of the existing mathematical knowledge into a clear, uniform pattern. He arranged everything in brilliant order, and knit it together with rigorous deductive reasoning.

More than 1000 editions of Euclid's *Elements* have been published. No other book—except the Bible—has been translated, edited, or studied more. For over 2000 years, Euclid's work was the standard textbook of mathematics; its influence on scientific thinking has never been equalled.

The Search for a Royal Road

King Ptolemy not only supported education for his people, but he also wanted to learn. After listening to Euclid explain some of the basics of geometry, he interrupted.

"Surely," the king complained, "surely, there is an easier way! I don't have time to learn all of this."

Euclid boldly responded. "My King, in the real world there are two roads—one for the common people and one for the King. But in geometry, there is no royal road."

The Impatient Student

The ancient scholar, Stobaeus, tells of a student studying geometry under Euclid. Like many modern students, this young man wanted to know what he would get out of learning mathematics. Euclid, in frustration, ordered his servant to give the student some copper coins, "since he must gain from what he learns!"

Training for a President

When Abraham Lincoln was 40 years old and struggling to make a living as a lawyer, he began to study Euclid's *Elements*. He was determined to discipline his mind for logical thinking. By mastering the first six books of Euclid, Lincoln sharpened his capacity for exact reasoning.

Warning: "Do Not Enter"

A thorough understanding of Euclid's work quickly became a prerequisite to advanced study in science and mathematics. Some Greek schools declared this by painting or engraving in the arches over their doors: "Let no one ignorant of geometry enter here."

Newton Changes His Mind

The first English translation of Euclid's *Elements* appeared in 1570, when William Shakespeare was only six years old. It took up 928 pages. Sir Isaac Newton first dismissed it as "trifling," but several years later he realized he had made a mistake. He went back and studied the work with fascination.

EUCLID'S ALGORITHM

What is the greatest common divisor of 1541 and 1943?
I found an easy method (algorithm) to do this! It's named after me.

Euclid

EXAMPLE: Find the GCD of 18 and 26.

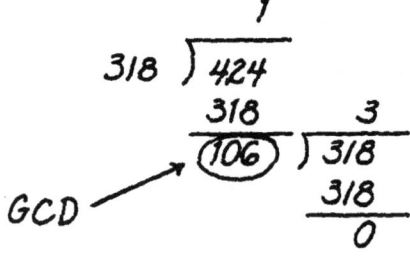

1.
2.
3.
4. GCD

EXAMPLE: Find the GCD of 318 and 424.

```
        1
318 ) 424
      318     3
     (106) ) 318
             318
               0
```
GCD

EUCLID'S ALGORITHM
To find the greatest common divisor (GCD) of two whole numbers, use this process:

1. Divide the larger of the two numbers by the smaller one.

2. If there is a remainder, divide it into the divisor.

3. Continue dividing the last divisor by the last remainder until the remainder is 0.

4. The final divisor is the greatest common divisor (GCD) of the original pair of numbers.

EXERCISES:
1. Use Euclid's algorithm to find the GCD for each pair of numbers.
   (a) 24 and 82 (b) 78 and 195 (c) 84 and 180
   (d) 84 and 324 (e) 660 and 840 (f) 1541 and 1943

2. Use Euclid's algorithm to reduce each of the fractions to lowest terms.

   (a) $\frac{56}{161}$ (b) $\frac{183}{427}$ (c) $\frac{291}{485}$ (d) $\frac{1182}{2758}$

(Hint: First find the GCD for the numerator and denominator.)

THE GREATEST COMMON DIVISOR MEETS THE LEAST COMMON MULTIPLE

The greatest common divisor (GCD) of two positive integers is the largest number that divides each of them.

The least common multiple (LCM) of two positive integers is the smallest number that is a multiple of each of them.

For example, consider the two numbers 12 and 15. The GCD is 3 and the LCM is 60.

Discover the relationship between any two positive integers and their GCD and LCM. Complete the table to help you.

A	B	GCD	LCM
8	12	4	24
6	40		
9	30		
24	36		
15	25		
18	4		

Use what you discovered above to complete the table.

A	B	GCD	LCM
33		11	66
	50	10	600
38		2	1064
	280	20	8680

Describe what you discovered about the relationship between A, B, GCD, and LCM.

HISTORICAL CONNECTIONS VOL. II

PERFECT NUMBERS

> Euclid wrote about perfect numbers in 300 B. C.
> A perfect number is a number which is equal to the sum of its proper divisors.
> 6 is a perfect number because 1 + 2 + 3 = 6.

Use the pattern you see to complete the first two columns in this table. Then, if the number in the second column is a prime number, multiply it times the number in the first column.

The result is a perfect number!

The early Greeks were fascinated by perfect numbers, but they could find only four. After completing this activity, you will have found the first five perfect numbers.

Perfectly amazing facts about perfect numbers

- In 1992, only 32 perfect numbers were known. The 32nd was found using the world's most powerful computer. It has 455,663 digits!
- No one has found an odd perfect number.
- All even perfect numbers end in 6 or 8.
- All even perfect numbers are also triangular numbers.

1st Column	2nd Column	Perfect Numbers
2	3	6
4	7	
8	15	
16	31	
32	63	

HISTORICAL CONNECTIONS VOL. II

DAFFY DEFINITIONS FROM GEOMETRY

Match the geometric terms in the box with these "daffy" definitions:

_____ 1. An angle that is never wrong.

_____ 2. A blending of musical tones.

_____ 3. A flattering remark.

_____ 4. What Noah built.

_____ 5. What bloodhounds do in chasing down a woman criminal.

_____ 6. A measure of temperature.

_____ 7. A good looking angle.

_____ 8. A missing parrot.

_____ 9. The way a poet writes love letters.

_____ 10. It's good to eat.

_____ 11. They voted yes on farm machinery.

_____ 12. What some girls want to find at the beach.

_____ 13. A ruined angle.

_____ 14. The one in charge.

_____ 15. What little acorns say when they grow up.

_____ 16. A tall coffee pot in use.

_____ 17. What one puts ice cream in at a picnic.

_____ 18. A gram, but not a measure of weight.

_____ 19. A way of catching a weird animal.

_____ 20. The name of a show with Mr. Ray as manager.

_____ 21. What you say to a "Sir" who is in the ant control business.

_____ 22. King of the jungle.

_____ 23. What Orville and Wilbur built.

_____ 24. What a person should do when it rains.

_____ 25. A state of blindness.

Acute angle
Arc
Center
Chord
Circumference
Coincide
Complement
Cone
Degree
Geometry
Hypotenuse
Inverse
Line
Parallelogram
Pi
Plane
Polygon
Protractor
Ratio
Rectangle
Right angle
Ruler
Secant
Tangent
Trapezoid

WHITE-FACED CUBES

 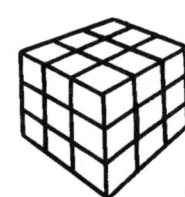  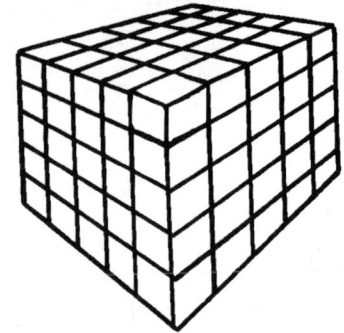

Suppose wooden cubes of various sizes are painted white and then cut into smaller cubes as shown above.

How many smaller cubes have 3 white faces? 2 white faces? 1 white face? no white faces?

Complete the table and discover the many patterns that emerge.

Length of edge	Number of cubes formed	Number with 3 white faces	Number with 2 white faces	Number with 1 white face	Number with no white faces
2					
3					
4					
5					
6					
20					
n					

HISTORICAL CONNECTIONS VOL. II © 2005 AIMS Education Foundation

Heron
c. 75

HERON
THE INGENIOUS ENGINEER

Biographical Facts:

Heron of Alexandria was a Greek mathematician, physicist, and engineer who lived about 75 A.D. Very little is known about his personal life. Heron (or Hero) wrote several treatises that reveal a remarkable grasp of geometry, optics, pneumatics, and mechanics.

Contributions:

Heron is best known for his formula for finding the area of a triangle. He also described an amazingly accurate method for finding the square roots of numbers. Heron, whose interests were more practical than theoretical, did much to provide a scientific foundation for engineering and land surveying. He designed many ingenious mechanical contrivances, including toys, a fire engine, a device for opening temple doors by a fire on the altar, and a wind organ.

Anecdotes:

Engineering Ingenuity

Heron's book, *Pneumatica*, describes more than one hundred machines and toys. Some of these were clearly devised just for fun, but others had practical applications. Heron's study of gases and liquids led to the development of a water-driven organ, a gun powered by compressed air, and a hose for spraying liquid fire. His work with steam enabled him to design public gardens with water fountains and moving statues driven by water pressure. There are even reports of "automobiles" being powered by steam along the city streets in annual religious parades.

In *Metrica*, his chief geometrical work, Heron applied his knowledge of trigonometry to describe how to dig a tunnel through a mountain by excavating simultaneously at each end.

Letting Off Steam

Heron's engine is one of the earliest illustrations of the principles behind jet propulsion. When water was made to boil in a hollow ball, the steam emerging from jets on the two sides propelled the ball into high speed rotation. This is because the force on the steam driving it out of the ball was accompanied by an equal and opposite reaction force exerted back on the ball by the emerging steam, as Newton explained later.

The Temple "Miracles"

During the first century, devout Greeks and Egyptians were sometimes awestruck when they entered their temples for worship. As people placed their coins in the temple coffers, selected stone gods shed tears, raised their hands to bless the worshippers, or poured out libations. In some temples, mechanical doves rose and descended. Heron both created and explained some of these "miracles." Most of them occurred through the use of hidden contraptions which used steam and a series of pulleys and gears.

Temple doors which opened automatically were among the most stunning mysteries. As worshippers approached, the doors silently glided open, revealing the fire on the altar within. Great ingenuity went into the construction of the opening doors. When the fire was lit on the altar, the air in the space beneath it expanded. This hot air was forced into a water filled globe, causing the water to overflow into a pail. This pail was connected by a cable to the doors. As the pail became heavy, the doors mysteriously opened.

Mirror, Mirror, On The Wall

Heron devoted an entire book, *Catoptrics*, to the theory of mirrors and applications of that theory. Although the study was completely scientific, Heron's results might have qualified him to work in an amusement park! Heron designed mirrors to see the back of the head and to reflect people standing upside down. Some of his creations even gave viewers extra eyes and noses.

Nature Solves the "Least Distance" Problem

Consider a modern version of a problem Heron solved. Suppose A and B are the positions of two towns and PQ is a river. A pier which will serve both towns is to be built along the river so that the total distance from the pier to town A and from the pier to town B is as short as possible. At what point, X, along the river should the pier be built?

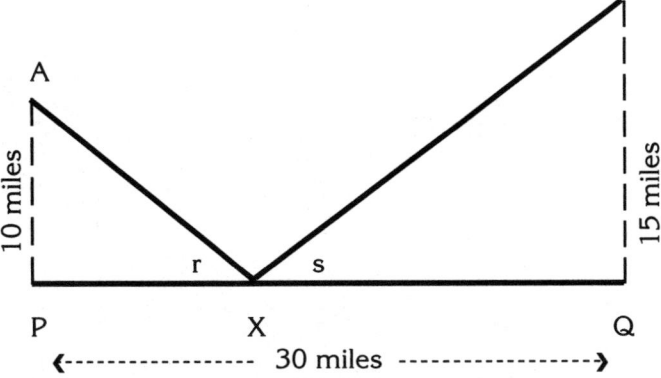

Heron believed Aristotle's principle that nature never does things the hard way. Heron imagined a light source at A, the eye of an observer at B, and a plane mirror along the river. The beam would take the shortest path from A to the river and then to B. This path results in ∠r being equal to ∠s since when light is reflected from a surface, the angle of incidence is equal to the angle of reflection. Therefore, point X must be found on the river so ∠r is equal to ∠s. Point X can be found by dropping perpendiculars from A and B to line PQ and using the resulting similar right triangles. The solution in our example is PX = 12 miles.

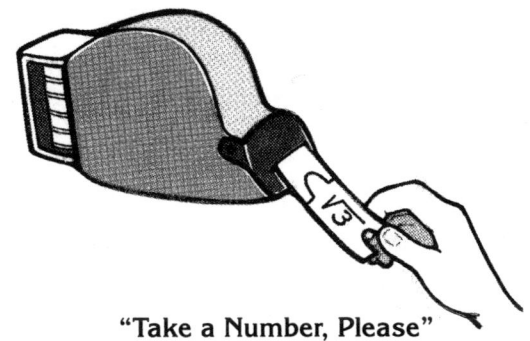

"Take a Number, Please"

Heron Finds Square Roots

Today it is easy to find the square root of a number by merely pushing a key on a pocket calculator. Before the days of computers and calculators, however, finding square roots was a little more difficult. As a practical engineer, Heron had to solve many problems involving square roots.

Heron's method, sometimes referred to as "divide and average," is one of the most efficient formulas ever devised for finding the square root of a number. Results are amazingly accurate after only two approximations; after five approximations, they may be accurate to 17 decimal places! The method is frequently used in computers and calculators today.

Heron's Formula

Heron's formula is an amazing formula since it allows us to compute the area of a triangle given only the length of the three sides. The altitude does not have to be known. Heron's formula has many practical applications. For example, a surveyor who knows the lengths of a three-sided lot can easily determine its area using Heron's formula. Parcels of land with four or more sides can be decomposed into triangular parts.

HERON'S SQUARE ROOT METHOD

As a practical engineer, Heron of Alexandria had to solve many problems involving square roots. His formula for finding square roots is one of the most efficient formulas ever devised. It is frequently used in computers and calculators.

HERON'S METHOD FOR FINDING SQUARE ROOTS
1) Approximate square root of desired number, N.
2) Find the average of your approximation and $\frac{N}{approximation}$. This is your new approximation.
3) Repeat step 2 until desired accuracy is obtained.

Study the examples below to understand the process.

Example 1: Find the square root of 8. Round to two decimal places.
First approximation: Choose 3 since $3^2 = 9$ is the closest square to 8.
Second approximation: $\frac{1}{2}(3 + \frac{8}{3}) = \frac{17}{6} = 2.83$

Example 2: Find the square root of 23.
First approximation: 5
Second approximation: $\frac{1}{2}(5 + \frac{23}{5}) = \frac{48}{10} = 4.80$

EXERCISES:

Find the approximate square root of each of these numbers by using Heron's Method. Use two approximations and round to the nearest hundredth.

Compare your results with square roots found using a calculator.

1. 37

2. 27

3. 14

4. 87

HERON'S FORMULA

Guess which of these two triangles has the larger area.

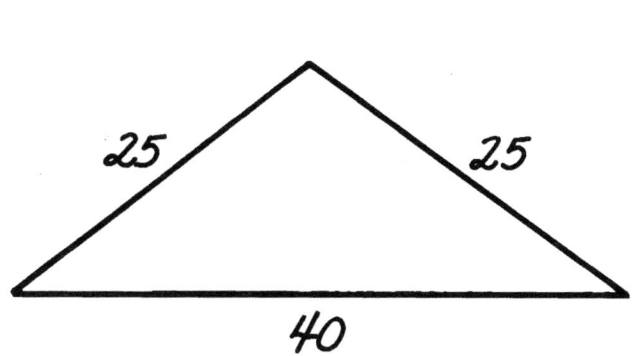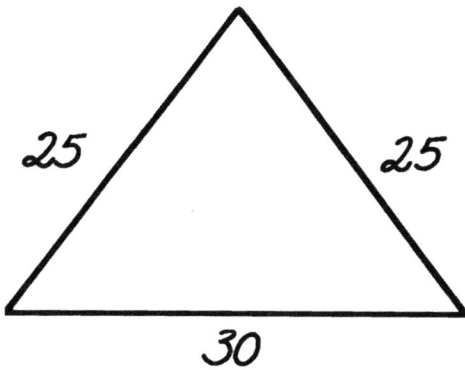

The area of each of the triangles above may be found by using a formula derived by Heron of Alexandria. The formula allows us to find the area of any triangle in terms of the length of the sides.

Heron's formula: $A = \sqrt{s(s-a)(s-b)(s-c)}$

where a, b, and c are the sides and $s = \frac{a+b+c}{2}$, the semiperimeter.

Example: Consider the triangle with sides, 6, 8, and 10. For this triangle $s = \frac{6+8+10}{2} = 12$. Hence, the area of this triangle is $\sqrt{12(12-6)(12-8)(12-10)} = \sqrt{(12)(6)(4)(2)} = \sqrt{576} = 24$.

EXERCISES:

1. Use Heron's formula to find the areas of the two triangles above. Which has the larger area? Are you surprised?

2. For the right triangle with sides 3, 4, and 5, find the area using
   (a) Heron's formula.
   (b) the formula $A = \frac{bh}{2}$.

3. Find the area of the triangle with sides 17, 25, and 26 using Heron's formula.

THE CHOCOLATE CAKE CHALLENGE

Paul is having a birthday party with 6 of his friends. His mother has baked a square chocolate cake, 7 inches on a side and 2 inches high. The cake has frosting uniformly spread on the top and sides.

Paul wants everyone at the party to receive the same amount of cake and the same amount of frosting, but his mother isn't sure how to cut the cake into 7 equal parts.

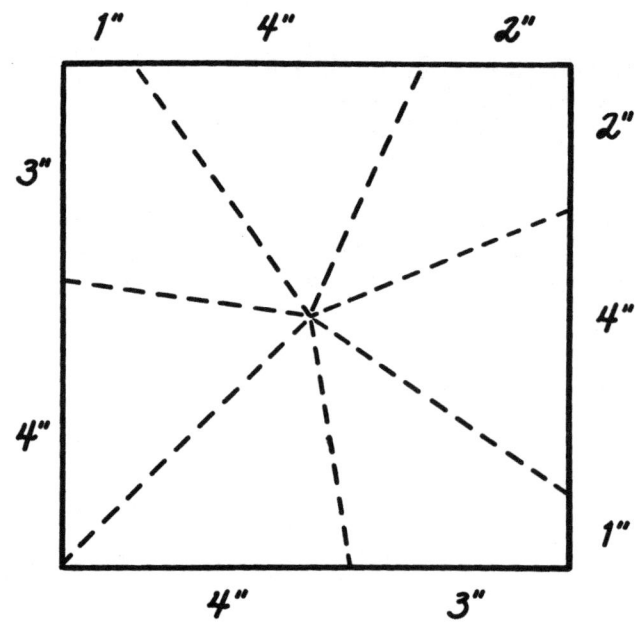

Max suggests that Paul's mother divide the perimeter (28 inches) into 7 equal parts and then cut from the center of the cake to the appropriate points on the perimeter.

Show that Max's solution guarantees that each of the 7 persons will get the same amount of cake and the same amount of frosting.

HISTORICAL CONNECTIONS VOL. II © 2005 AIMS Education Foundation

SPACE FOR ACE

Sometimes Mack takes his labrador with him to work at his auto body shop. He hooks Ace's collar to a 60 foot chain mounted on the corner of the shop, which is a 40 by 80 foot rectangular building.

Ace enjoys going to work with Mack; he gets to explore a lot more territory than he does in the family's back yard.

What is the area of the region Ace can enjoy at the shop?

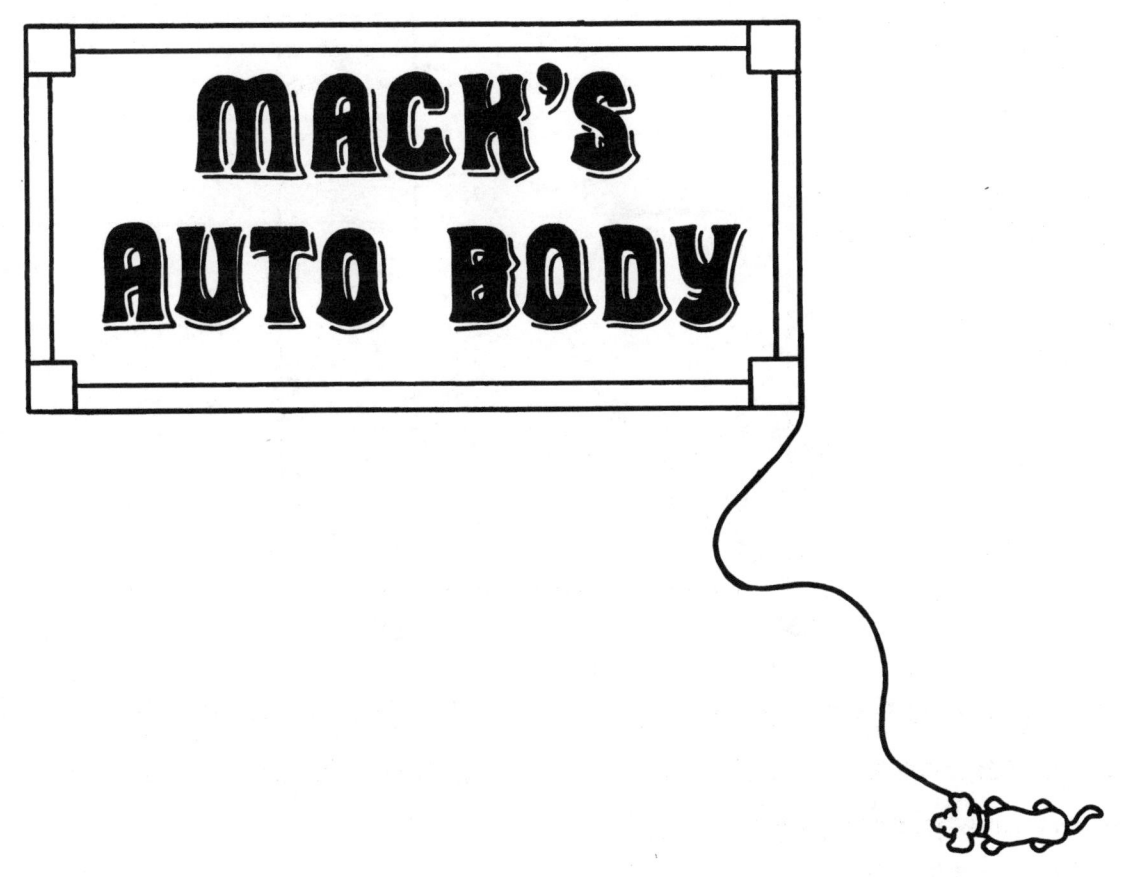

DIAGONAL CHALLENGE

The early Greeks were fascinated by geometry. Share this interest by predicting how many diagonals can be drawn in polygons with 8, 9, and 10 sides.

First, count the diagonals in a 3-sided polygon, a 4-sided polygon, and a 5-sided polygon. Count the diagonals as you draw them for the 6- sided and 7-sided polygons. When all possible diagonals have been drawn, the number drawn from each vertex should be the same.

Record the number of diagonals in the table. What pattern appears? Use this pattern to predict the answers for 8, 9, and 10-sided polygons without actually drawing them.

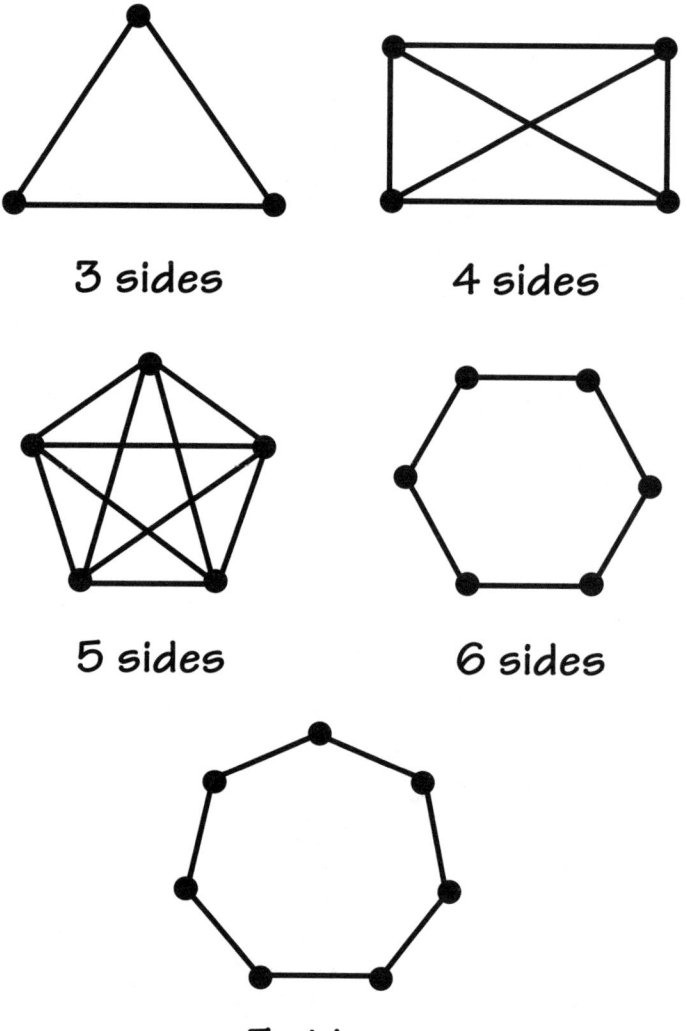

Number of Sides	Number of Diagonals
3	0
4	2
5	5
6	
7	
8	
9	
10	

Can you state a formula for finding the number of diagonals in a polygon with n sides?

HISTORICAL CONNECTIONS VOL. II   24   © 2005 AIMS Education Foundation

Hypatia
c. 370 - c. 415

HYPATIA
MODEL OF EXCELLENCE

Biographical Facts:

Hypatia (hy-PAY-shuh) was a brilliant Greek scholar who lived from about 370 to 415. She is the first known woman mathematician. Following in the footsteps of her father, Theon, she became a professor at the University of Alexandria. Her cruel death marked the end of the impressive Greek era.

Contributions:

One of Hypatia's greatest gifts was her ability to explain complicated concepts in simple ways. She composed texts about Diophantine algebra, the Conics of Apollonius, and the works of Ptolemy and Euclid, explaining and illustrating for the benefit of her students. Though she was primarily an algebraist, she also invented and built several scientific instruments, including an astrolabe and a hydrometer.

Anecdotes:

Like Father, Like Daughter

As a child, Hypatia spent most of her time with her father Theon, distinguished professor of mathematics at the University of Alexandria. Learning, questioning, and exploring became as natural to her as breathing and eating. Hypatia caught Theon's love of mathematics as he taught and tutored her. But he wanted her to be well-rounded, so he scheduled time for play and physical fitness activities like running and climbing. He believed that the whole person—body, mind, and spirit—should be developed. As a result, Hypatia became a beautiful woman, gifted in public speaking, and trained in the arts, literature, science, and philosophy.

A Stimulating Environment

Hypatia had the advantage of growing up in Alexandria, a cosmopolitan center where the finest scholars gathered to exchange ideas. Because of its lovely setting on the Mediterranean Sea, the city had been chosen by Ptolemy in 323 B.C. to be the capital of his kingdom and the site of the first university. By the time Hypatia was born, the university was nearly 700 years old, and a series of invasions by Roman armies had weakened Greek tradition. Hypatia was one of the last great teachers to uphold the Greek values of creative inquiry.

In Love With Her Work

Hypatia was never happier than in the classroom, teaching and sharing ideas with her students. The historian Socrates said that her home and lecture room were the two favorite spots of scholars. Occasionally, she abandoned the formal classroom for the center of the city. There, among the city officials and the common folk, she carried on discussions that inspired all who listened. Some suspected that she was an *oracle;* students came from Europe, Asia, and Africa to sit under her instruction.

Accused of Heresy

Theon taught his daughter to be open-minded about religion. No belief should be final, he said, and one should always examine ideas and seek the truth. Although Christianity gained many followers during her lifetime,

Hypatia remained loyal to the Greek religion. When Cyril became archbishop of Alexandria in 412, Hypatia was quickly labeled a heretic. Cyril's followers became jealous of the large crowds that gathered to hear Hypatia teach. Soon they began to spread rumors that Hypatia was teaching paganism under the cover of teaching mathematics and science.

A Deadly Intrigue

One of Hypatia's friends was Orestes, the Jewish prefect of Egypt. As antagonism between Orestes and Cyril escalated, Hypatia became caught in the middle. At first there were just small incidents of harassment between the two political/religious groups, but when diplomatic attempts to resolve the conflict were unsuccessful, Cyril became outraged. He and his supporters believed that Hypatia was keeping Orestes from peaceful negotiation. Her death, they believed, would make a strong statement about their power. Orestes, sensing what was about to happen, begged Hypatia to withdraw her political stance, but she refused.

Brutal Murder

One day in 415, Hypatia was on her way to classes at the university when a mob of fanatics pulled her from her chariot and committed one of the most savage murders described in human history. They dragged her to the cathedral in Alexandria, stripped her naked, and slashed her with sharp oyster shells. One account says that her attackers pulled out her hair. Her body was cut into pieces and burned.

No Justice in Rome

Orestes felt deep pain and responsibility for Hypatia's fate; he tried to get Roman officials to investigate. The officials said there was an unfortunate "lack of witnesses." Finally, the Bishop announced that an investigation was unnecessary because Hypatia was alive and well in Athens. No tragedy had occurred, he falsely asserted. Yet the development of mathematics fell into a period of decline lasting 1000 years.

THE AMAZING ELLIPSE

When John Quincy Adams was a congressman, the House of Representatives met in Statuary Hall, in the Capitol building. On several occasions, Adams amazed Congress by seeming to know exactly what his political opponents were going to do before they did it. How did he find out?

Statuary Hall is shaped like an ellipse. Whispers from a focus point in such rooms bounce off the wall and reach the other focus point (in this case, Adams' desk) in a concentrated form.

In 1857, because of this acoustic phenomenon, the House chose to meet elsewhere. On tours of the Capitol, the "whispering" property of the ellipse can still be demonstrated.

DRAWING ELLIPSES:
- Tack a sheet of graph paper on cardboard.
- Tie a loop at each end of a piece of string so that it is about 8 inches long.
- Put the loops on two tacks or push pins at the focus points marked A.
- Using your pencil, pull the string taut and move the pencil around the paper to draw an ellipse.
- Draw two additional ellipses using the pair of points marked B and the pair of points marked C.
- What happens to the shape of the ellipse as the points get closer to the center?

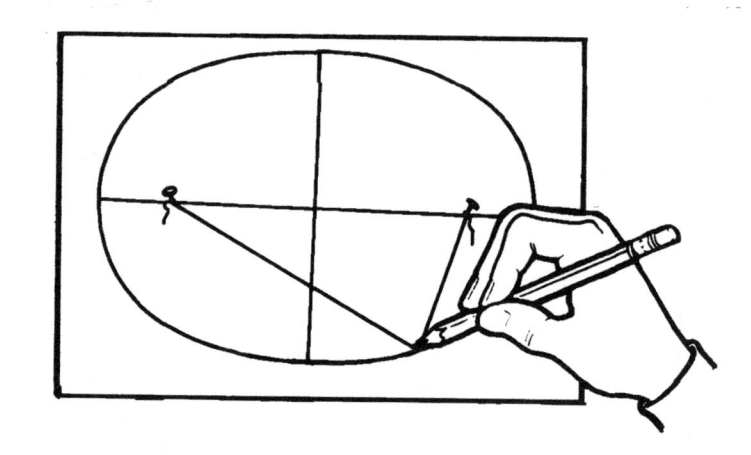

DEFINITION: An ellipse is the set of all points in a plane such that the sum of distances (in this case, the length of the string) to two fixed points is constant. Each fixed point is called a focus.

HISTORICAL FACTS:

Hypatia (370 - 415), the first known woman mathematician, wrote about ellipses and other conic sections.

In 1609, Johann Kepler established the importance of the ellipse when he discovered that the orbits of the planets around the sun are ellipses and not circles as the ancients thought.

The Amazing Ellipse

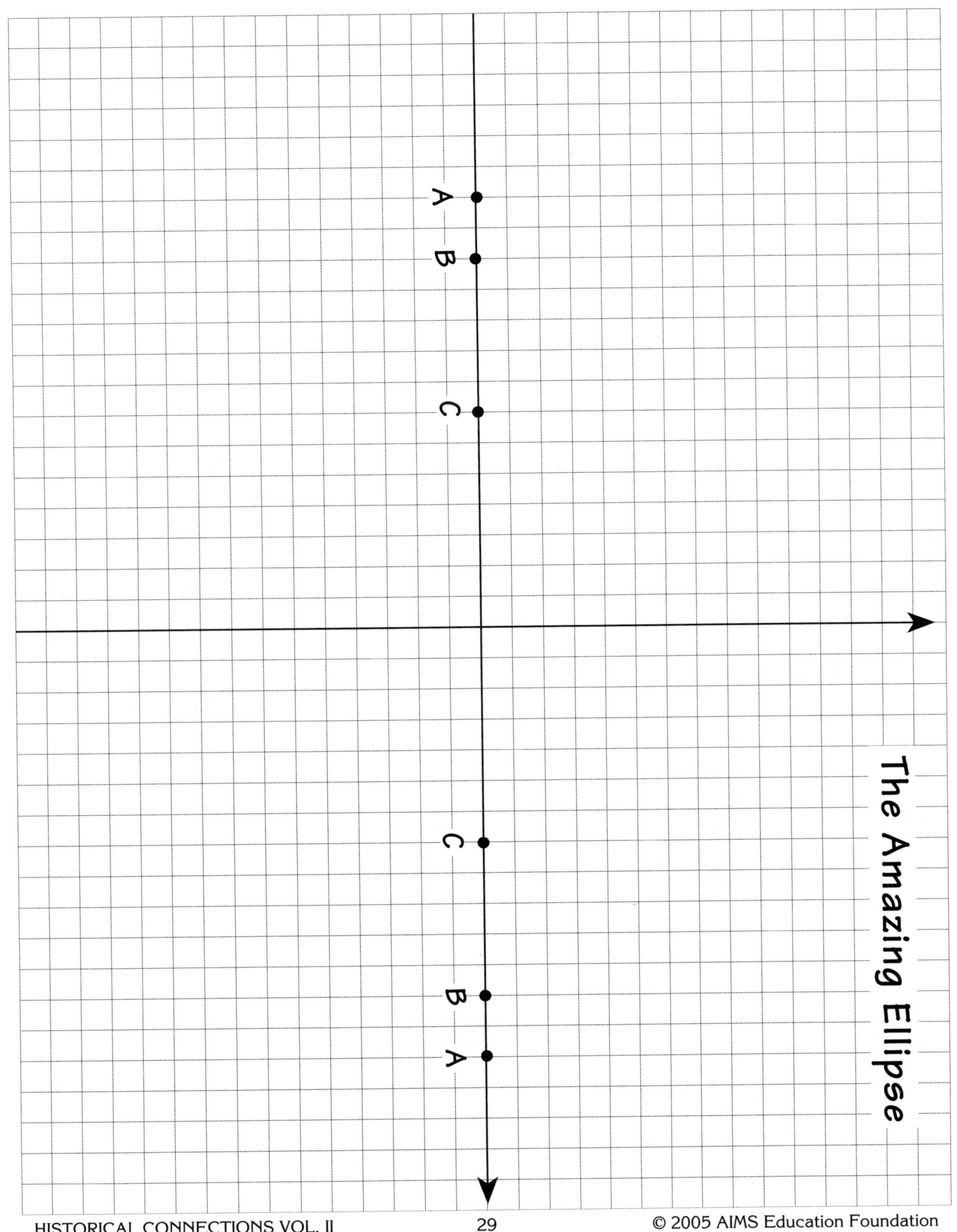

HISTORICAL CONNECTIONS VOL. II © 2005 AIMS Education Foundation

CHANGING A QUARTER

Kia and her brother, Max, wanted to buy gumballs from the machine in the grocery store, but all they had was a quarter, which the machine would not take.

"Could we please have change for this quarter?" they asked the clerk. The clerk, tired of making change for kids, made them an offer instead.

"If you can tell me in how many ways change can be made for a quarter," he chuckled, "I'll give you each a free gumball!"

Help Kia and Max solve this problem by completing the table.

Number of Dimes	Number of Nickels	Number of Pennies
2	1	0

Add as many rows to the table as you need to solve this problem.

SUPER CHALLENGE:

In how many ways can you make change for one dollar? You may use any combination of pennies, nickels, dimes, quarters, and half-dollars.

PAPER FOLDING A PARABOLA

Hypatia, the first woman mathematician, wrote about the parabola, a special conic section. In this activity you will have an opportunity to create a parabola by folding paper.

First, cut off the lower half of this activity sheet. Fold the paper so that the lower edge touches the point marked near the bottom. Make a crease with your fingernail, open the paper, and fold it again at a different angle, always making sure the bottom edge of the paper touches the point. Repeat this about twenty times. A parabola should appear, which you may outline with a pencil.

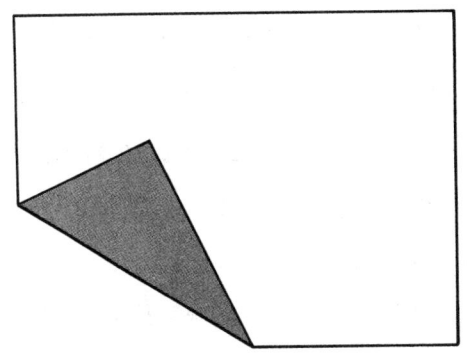

Try this activity again with a larger sheet of paper. Mark a point in the center, about one inch above the lower edge.

Galileo showed that the path of an object thrown through space is a parabola. When a ball is hit into the air, it follows a path in the shape of a curve called a parabola.

Parabolic mirrors are used in automobile headlights and (on a larger scale) in searchlights. Parabolic reflectors are also used in some telescopes and antennas to collect light and radio waves from outer space.

A STAMP STUMPER

Here is a modern day version of the kind of problem Hypatia, the first woman mathematician, wrote about in 400. These were known as Diophantine problems.

Marcia has an abundant supply of 3-cent stamps and 8-cent stamps. She needs to mail a package that requires exactly $2.00 postage. How can she come up with exactly $2.00 using any combination of the 3-cent and 8-cent stamps?

There are 9 different solutions to this problem. Try to find them all and list them in this table.

Number of 3-cent stamps	Number of 8-cent stamps

Did you observe any patterns?

Describe the strategy you used to solve this problem.

Banneker
1731 - 1806

BENJAMIN BANNEKER
SELF-TAUGHT GENIUS

Biographical Facts:

Benjamin Banneker was a free African-American tobacco farmer who achieved fame as a compiler of almanacs. He was born near Ellicott's Lower Mills, Maryland, on November 9, 1731. His grandmother was an indentured sevant from Enland who bought his Ethiopian grandfather's freedom. Banneker lived most of his life on his farm in Baltimore County, and died October 9, 1806.

Contributions:

Benjamin Banneker is a striking example of the ability people have to succeed despite inadequate materials and educational opportunities. Banneker's almanacs (1791-1797) were widely distributed to counter the prevailing opinion that African-Americans were intellectually inferior to whites. He was an assistant in astronomy during the surveying of the boundaries for the new District of Columbia.

Quotation by Banneker:

"The color of the skin is in no way connected with the strength of the mind or intellectual powers."

Anecdotes:

Counting the Steps

As soon as he was old enough to help, Benjamin Banneker spent most of his time cultivating tobacco on his family's farm. The process was time-consuming and demanding. The tiny seeds were first planted in a seedbed in the woods. About a month later, the new plants were carefully transplanted. The rows of young plants had to be watched closely for insects and worms, which were picked off by hand. The family spent the next months pruning and topping the plants, removing suckers, and hoeing weeds. Finally, it was time for harvest.

Benjamin shared all of the farming duties with his parents. Once, he calculated that raising a tobacco plant from seed to harvest required thirty-six separate operations. Even as a child, he was interested in counting and keeping records.

Time for School

While tobacco farming was not particularly hard work, it required continuous attention except for the very cold winter months. During this season, Benjamin was sent to a small one-room school near his home. The school, run by Quakers, served several white children as well as three or four African-American children. Banneker's grandmother had taught him to read and write, but at school he also learned arithmetic, which quickly became his favorite subject. As he grew older, the family needed him on the farm, so he had only several years of formal education.

The Striking Clock

The Banneker family had no clocks or timepieces. Like most of their neighbors, they planned their days around the sunrise and sunset. But one day Banneker had an opportunity to borrow a pocket watch. After observing its intricate mechanism, he set out to build one for himself. He drew sketches of the various wheels and gears, and applied his mathematical skill to understand how it all fit together and worked.

Over a long period of time, Banneker worked on his clock, carving and whittling each piece out of hard-grained wood. Finally, he assembled the many parts. The clock worked beautifully! The large striking clock quickly became the talk of the community. People who had never heard of Banneker came from miles around to observe his masterpiece. He was only 22 years old.

An Independent Learner

Although Banneker had no books of his own, he determined to keep learning. In the few hours of free time he had each day, he read and studied books he borrowed. He was also an avid observer of nature, and paid close attention to the patterns in the seasons and in the plants and wildlife around him. It was an exciting day for Banneker when he bought his first book—a Bible—at the age of 32.

Contentment in Solitude

Banneker never married, and he had few close friends, but there is no evidence that he was lonely. He loved to go for long walks in the country, and tended his garden and bee hives with patient care. Someone gave him an old flute and a violin, and he learned to play them both. Neighbors passing by would often hear him making music in the evenings, outside under a favorite fruit tree. He loved puzzles of all kinds, especially those that required algebra to solve. He often copied puzzles about farm animals and household objects into his journal so that he could enjoy them again and again.

Influential Neighbors

Banneker's opportunities for learning multiplied when the Ellicott family moved into the area and built their gristmills. Banneker was fascinated by the mechanization their mills utilized. He took every chance he got to observe their equipment and listen to their conversation.

His friendship with George Ellicott was especially valuable. Ellicott was almost 30 years younger than Banneker, but their shared interests in astronomy and science drew them together. Eventually, Ellicott lent Banneker globes, books, and telescopes, promising to come back and show him how to use it all when he had time. But when Ellicott came back several months later, Banneker had mastered it all on his own.

Indispensable Almanacs

In colonial American family life, an almanac was an almost indispensable source of information. It was the only way to know what time the sun would rise and set on a given day. Farmers relied on it for weather forecasts and planting advice. Navigators needed it for schedules of tides and positions of fixed stars. Everyone depended on it as a source of entertainment and knowledge because it contained literary and historical items as well as more practical information.

The *ephemeris* was the most crucial section of the almanac, and the most difficult to compile. This collection of tables showing the positions and motions of the sun, moon, and stars from day to day had to be assembled with excruciating accuracy. The work demanded thousands of careful computations, and had to be done anew for each year.

More for his own pleasure than with any thought of publication, Banneker began to put together an ephemeris. When George Ellicott saw Banneker's work, he encouraged him to submit it for publication. Banneker had found a way to make a unique contribution to his society!

An Overnight Success

Banneker's almanacs were published from 1791 to 1797, and widely distributed throughout Pennsylvania, Maryland, Delaware, and Virginia. They were recognized for their accuracy and value, but also because they proved that an African-American could be gifted with scientific and mathematical abilities. Several of the Societies for the Promotion of the Abolition of Slavery sold the almanacs and used them in their arguments against slavery.

Thomas Jefferson, then Secretary of State, corresponded with Banneker and sent a copy of his first published almanac to the Marquis de Condorcet, Secretary of the Royal Academy of Sciences in Paris, to "show the talents of the Negro race."

Surveying the Future Capital

When Major Andrew Ellicott was commissioned to survey a ten-mile square section of land to become the nation's new capital, he looked for qualified assistants. Surveying required laborious computations based on an astronomical clock. A small miscalculation could cause a major error.

Ellicott chose Benjamin Banneker to be one of his assistants. Called to help survey the Federal Territory (the future Washington, D. C.), Banneker left his farm for the first and only time in his life. He was fascinated with the equipment at his disposal, and he did his assignment with dedication. In his honor, a Washington, D.C. park is named "Ben Banneker Park."

From Ashes to Ashes

Benjamin Banneker died in 1806, just a month before his 75th birthday. He had been in poor health for some time and had given many of his possessions to his sisters and their children. At the moment his body was lowered into the grave during the burial service on the family farm, his house mysteriously burst into flames. Everything he owned, including some manuscripts and the amazing striking clock, was destroyed. But his legacy, and his inspiring example of determination and dedication, lives on even today.

TWO PUZZLES
from
BENJAMIN BANNEKER'S COLLECTION

Benjamin Banneker, the first African-American mathematician, worked hard to compile a series of important almanacs—while also maintaining a farm. To relax, he collected and created mathematical puzzles. Here are two adapted from his collection.

THE LADDER PUZZLE

Suppose a ladder 50 feet long is placed in a street so that it reaches the roof of a building 48 feet tall. Without moving it at the bottom, it will reach the roof of a building 40 feet tall on the other side of the street. How wide is the street?

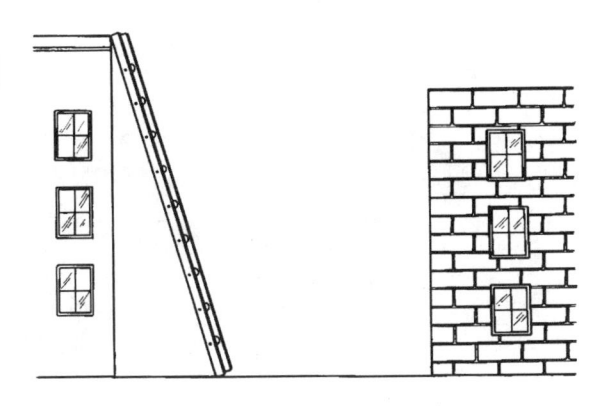

THE COWS, GOATS, AND CHICKENS PUZZLE

A gentleman farmer sent his servant with $100 to buy 100 animals, with orders to pay $5 for each cow, $1 for each goat, and 5¢ for each chicken. How many of each animal could the servant buy if he buys at least one of each?

A CLOCKMAKER'S CHALLENGE

Benjamin Banneker, an early African-American mathematician, amazed his friends and neighbors by building a wooden striking clock. He carefully carved the gears from wood. His only knowledge of how clocks worked came from studying a small borrowed pocket watch.

This activity explores some of the mathematics used in clockmaking.

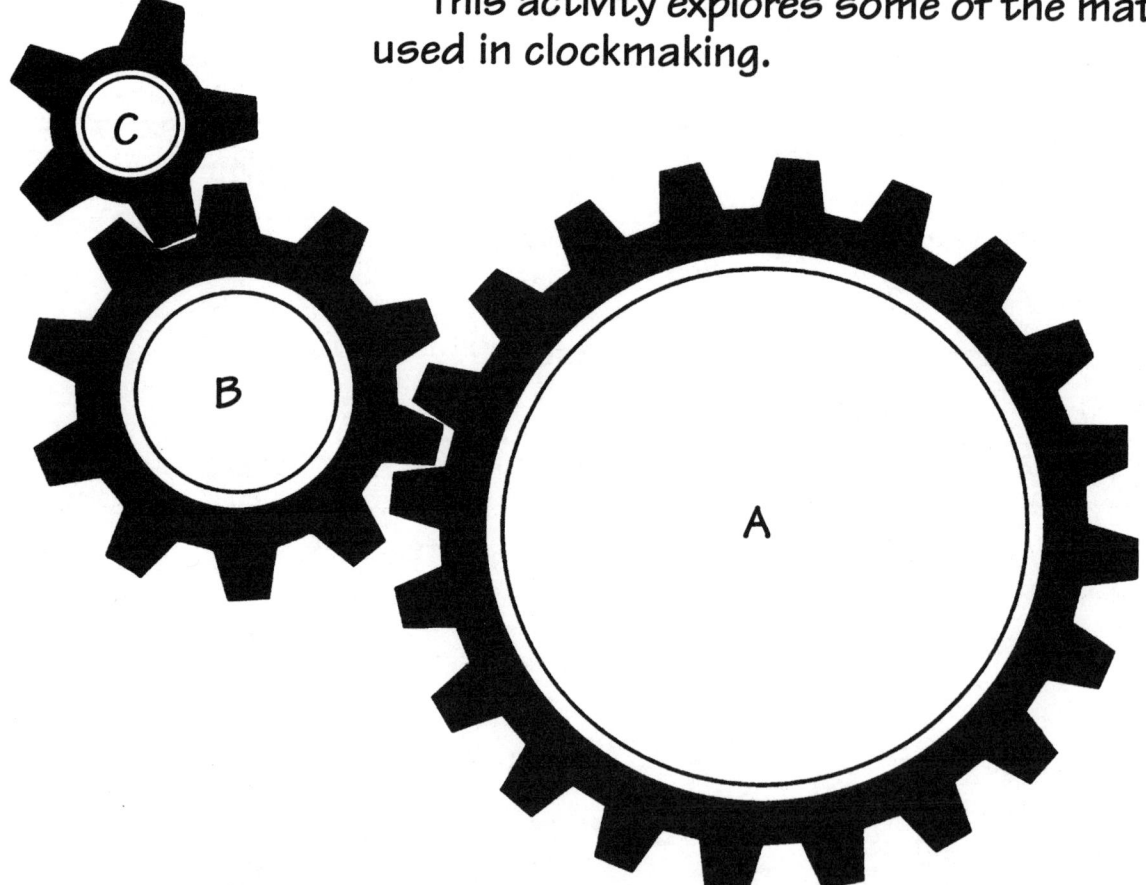

1. Suppose gear A makes one complete revolution clockwise.
   a. Will gear B move clockwise or counter-clockwise?
   b. Will gear C move clockwise or counter-clockwise?
   c. How many revolutions will gear B make?
   d. How many revolutions will gear C make?

2. Suppose gear B moves 8 revolutions clockwise.
   a. Will gear A move clockwise or counter-clockwise?
   b. How many revolutions will gear A make?
   c. How many revolutions will gear C make?

A COLLECTION OF PUZZLERS

Benjamin Banneker, the first African-American mathematician, enjoyed solving and collecting mathematical puzzles. Here are some challenging puzzlers for you to try.

1. The number of eggs in a basket doubles every minute. The basket is full in one hour. When was the basket half-full?

2. A bookworm starts at the outside of the front cover of Volume I of a set of five volumes of books on a bookshelf, and eats his way to the outside of the back cover of Volume V. If each of the volumes is one inch thick, how far has the bookworm eaten?

3. Two brothers, Josh and Ben, look exactly alike and are exactly the same age, yet they are not twins. How is this possible?

4. A frog is at the bottom of a 30 foot well. Each hour it climbs up 3 feet and slips back 2 feet. How many hours does it take the frog to get out?

5. Two cyclists are 10 miles apart and pedaling toward each other. One cyclist's rate is 6 miles per hour and the other's rate is 4 miles per hour. There is also a friendly fly zooming continuously back and forth from one bike to the other. If the fly's rate is 20 miles per hour, how far does the fly fly by the time the cyclists reach each other?

6. What is the minimum amount it will cost to make five three-link pieces of chain into one straight chain, if it costs 2 cents to cut a link and 2 cents to weld a link?

7. An artist once pointed to a painting he had painted and said:
   "Brothers and sisters have I none, but that man's father is my father's son."
   Whom had the artist painted?

8. If there are 10 blue socks and 16 grey socks in your drawer, and you reach into it in the dark, how many socks must you take out to be sure of getting a pair that matches?

9. "Johnson's Cat"
   Johnson's cat went up a tree,
   Which was sixty feet and three;
   Every day she climbed eleven,
   Every night she came down seven.
   Tell me, if she did not drop,
   When her paws would touch the top.

10. A squirrel goes spirally up a cylindrical post, making a circuit in each 4 feet. How many feet does it travel if the post is 16 feet high and 3 feet in circumference?

MAKE-A-SQUARE PUZZLE

Benjamin Banneker enjoyed collecting puzzles like this one.

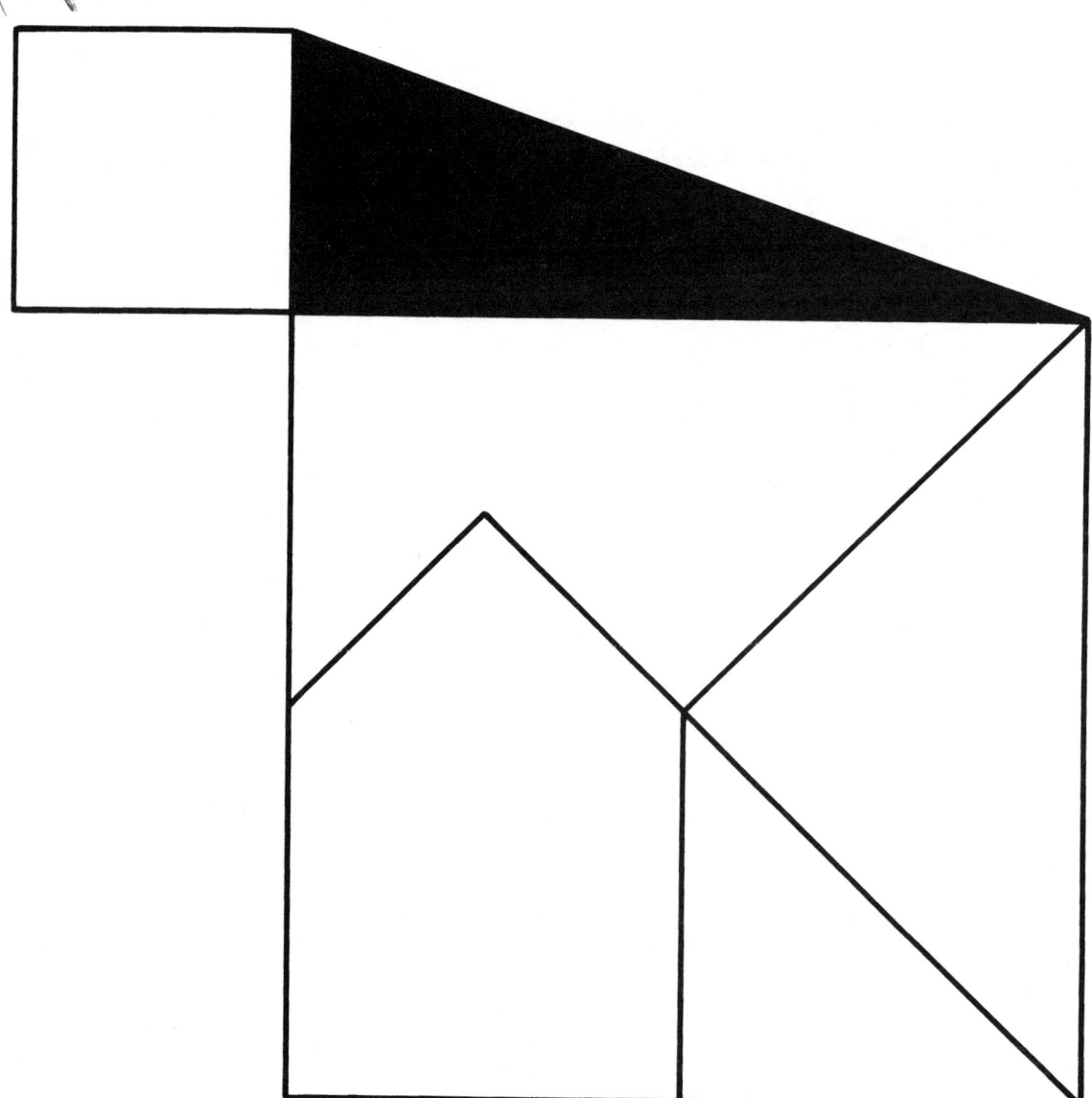

Cut out this puzzle and arrange the five unshaded pieces to form one large square on the hypotenuse of the shaded right triangle.

What famous theorem from geometry does this puzzle illustrate?

HISTORICAL CONNECTIONS VOL. II © 2005 AIMS Education Foundation

A JUMBLE PUZZLE

Unscramble these Jumbles, one letter to each square, to form five common words associated with Benjamin Banneker, the first African-American mathematician.

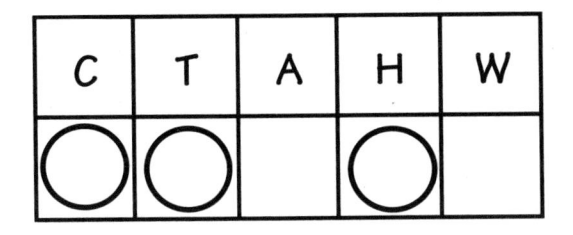

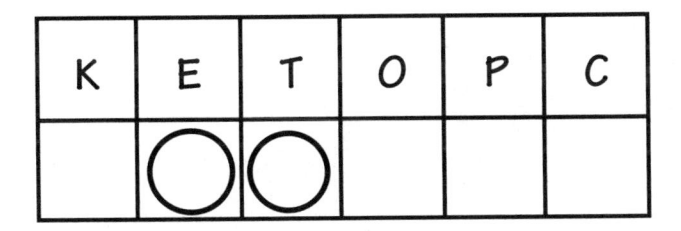

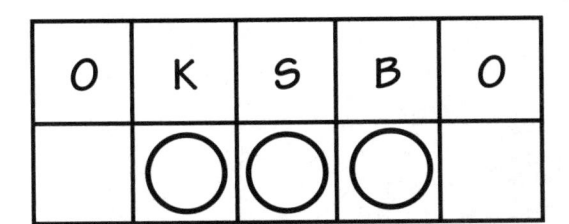

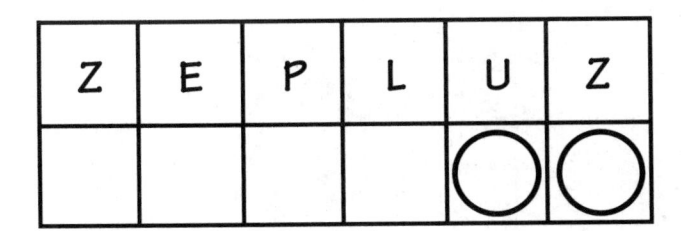

| C | T | A | H | W |

| K | E | T | O | P | C |

| O | K | S | B | O |

| Z | E | P | L | U | Z |

| L | Y | R | M | D | A | N | A |

What 'cha whittlin' Ben?

What Banneker got "geared up" for

Now arrange the circled letters to form the answer suggested by the above cartoon.

HISTORICAL CONNECTIONS VOL. II 41 © 2005 AIMS Education Foundation

Babbage
1792 - 1871

BABBAGE
MAKING A DIFFERENCE

Biographical Facts:
Charles Babbage was a British mathematician, engineer, and inventor who lived from 1792 to 1871. His father, a wealthy banker, left Babbage a large inheritance that freed him to pursue his mathematical interests.

Contributions:
Babbage's most important contribution to mathematics was his pioneering work in mechanical computation. His "engines," designed to perform many mathematical functions, are the clear forerunners of today's calculators and computers. He is also credited with inventing the speedometer, a cowcatcher for the British railroad, and a method for identifying lighthouses by their beams. His book, *Economy of Manufactures and Machinery*, "pre-visions" the field of operations research.

Anecdotes:

A Noble Pact
Babbage and his two best friends, John Hershel and George Peacock, attended Cambridge University together. While there, the three of them made a pact with each other: to "do their best to leave the world wiser than they found it."

The Difference Engine
During his student days at Cambridge, Babbage became frustrated with inaccuracies in the tables mathematics students were given to work with. He believed that tables of compound interest, logarithms, and trigonometric functions could be calculated more efficiently by machinery.

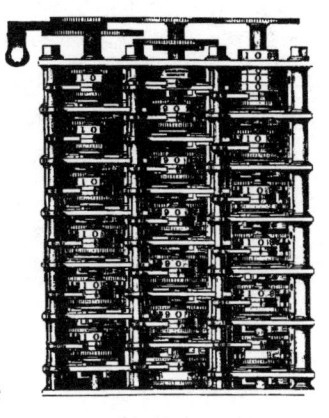

Babbage determined to build a machine that would produce error-free tables by utilizing differences. He envisioned a machine that, after the first few entries in a table were entered manually, would compute the differences between the entries and continue the pattern indefinitely.

Although the British government was persuaded to support the project, the work itself posed many challenges. New tools had to be created to manufacture parts for the machine Babbage designed. Construction was projected to take three years, but progress was slow. After 11 years and nearly $1,000,000 in government expenditures, the project was abandoned, but not because of technical difficulties. Babbage had a better idea!

Borrowing from a Weaver
J. M. Jacquard, a French weaver, invented the use of punched cards to control the sequence of threads in a loom. This technique created beautiful, intricate designs in woven carpets and fabrics. Babbage saw the looms in action and borrowed the idea of punched cards for his new project—the analytical engine. In appreciation, Babbage hung a colored portrait of Jacquard, woven in silk, on his wall. The portrait had been created on a loom using 20,000 punched cards.

The Analytical Engine
Imagine a machine which could add, multiply, divide, print, and call for new data from its storage bank. Such a piece of equipment would have seemed like science fiction in 1842, but this was Babbage's new dream. Unlike the difference engine, the analytical engine would use punched cards, as Babbage had seen used in carpet-making looms. Although Babbage devoted the rest of his life and his entire fortune to this venture, the machine was unfinished at his death. His son, H. P. Babbage, took up the cause from 1880

to 1910, but finally conceded that the plans were too complex for the engineering expertise of the day.

The Engine Takes a Census

Because it took the United States ten years to complete the 1880 census, the government sponsored a contest to find a more efficient method for counting the country's citizens. Herman Hollerith won the contest and was awarded the contract to conduct the 1890 count. Inspired by Babbage's designs, he built a machine using punched cards, counter wheels, and electromagnetic relays. While it was not capable of all the functions Babbage envisioned, it enabled Hollerith to complete the census in only one month.

Hollerith founded the Tabulating Machine Company which, in 1924, became the International Business Machines Corporation (IBM).

The IBM Connection

The most direct descendant of Babbage's analytical engine was the great IBM Sequence Controlled Calculator completed by Harvard and IBM in 1944 for the Navy. Another, the Electronic Numerical Integrator and Computer, was built in 1945 at the University of Pennsylvania for the Army. This machine requires a 30' X 30' room, contains 19,000 vacuum tubes, and weighs 30 tons. It is currently housed at the Smithsonian.

A Dream Comes True

In 1992, just in time for the 200th anniversary of Babbage's birth, a working model of the difference engine was completed by London's Science Museum, using only materials that would have been available to Babbage.

A consortium of computer companies contributed the $450,000 needed to build the computer, which stands 6 feet high, 10 feet long, and 18 inches deep.

Tennyson Corrected

Babbage demanded accuracy in everything he read and saw. In the poem, "The Vision of Sin," Alfred Lord Tennyson wrote,

"*Every minute dies a man,*
Every minute one is born."

Babbage wrote to Tennyson, suggesting he correct his "erroneous calculation." The second line, Babbage argued, should read "Every minute *one and a sixth* is born."

Tennyson responded by changing the word "minute" to "moment" in later editions.

Monkeying Around

The older Babbage got, the more crotchety he became. He once complained that he never had a happy day in his life! He seemed to find some satisfaction, however, in attacking his pet peeve: organ grinders. These "pests," he said, interfered with his concentration and destroyed his working power. His crusades against organ grinders made him notorious throughout London.

Ada Writes the Program

Ada Lovelace, daughter of British poet Lord Byron, was an aspiring young mathematician when she met Charles Babbage at a party. Babbage invited her and her mother to observe his work on the difference engine, thus beginning an 18-year relationship, largely through correspondence. Lovelace was intrigued by both the difference engine and the later analytical engine. She understood exactly what was going on. In fact, she was the first to write clear instructions on how to communicate with such machines. She provided the first examples of how a computer may be programmed. Her understanding of mathematics and her faith in Babbage, made concrete by financial support, was of invaluable help to him.

RINGS OF CIRCLES

Look what happens when circular objects (such as pennies or plastic chips) are arranged in rings around a central circle.

Count the circles in the designs with 3 and 4 rings and enter your results in the table. How many circles would be needed to create a design with 8 rings? Discover the pattern in the second column and complete the table.

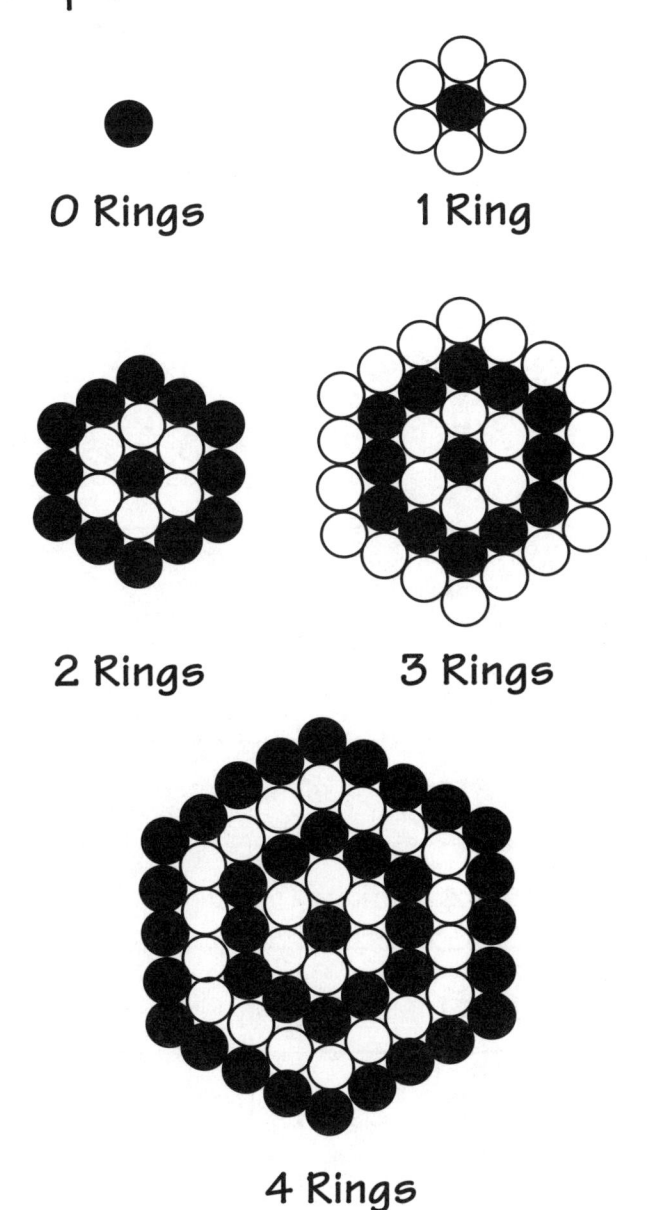

0 Rings

1 Ring

2 Rings

3 Rings

4 Rings

Number of Rings	Total Number of Circles
0	1
1	7
2	19
3	
4	
5	
6	
7	
8	

Charles Babbage (1792-1871) designed an amazing machine that would extend mathematical tables like this one indefinitely.

HISTORICAL CONNECTIONS VOL. II © 2005 AIMS Education Foundation

PIZZA PARTY

What is the maximum number of pieces possible when cutting a pizza with 8 straight cuts all the way across? The pieces do not need to be equal in size.

To solve this problem, count the pieces in pizzas with 1 cut, 2 cuts, 3 cuts, and 4 cuts.

Record the number of pieces, and use the pattern in that column to complete the table.

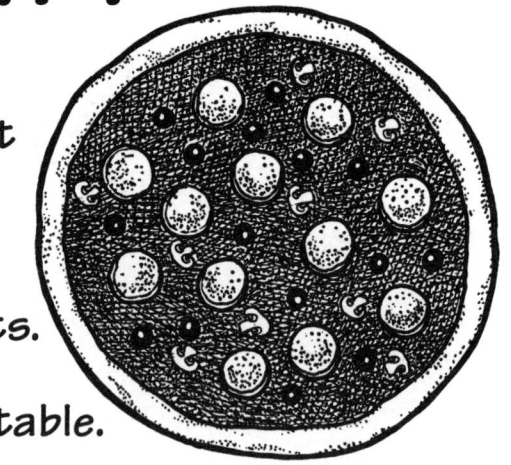

Can you develop a formula for n cuts?

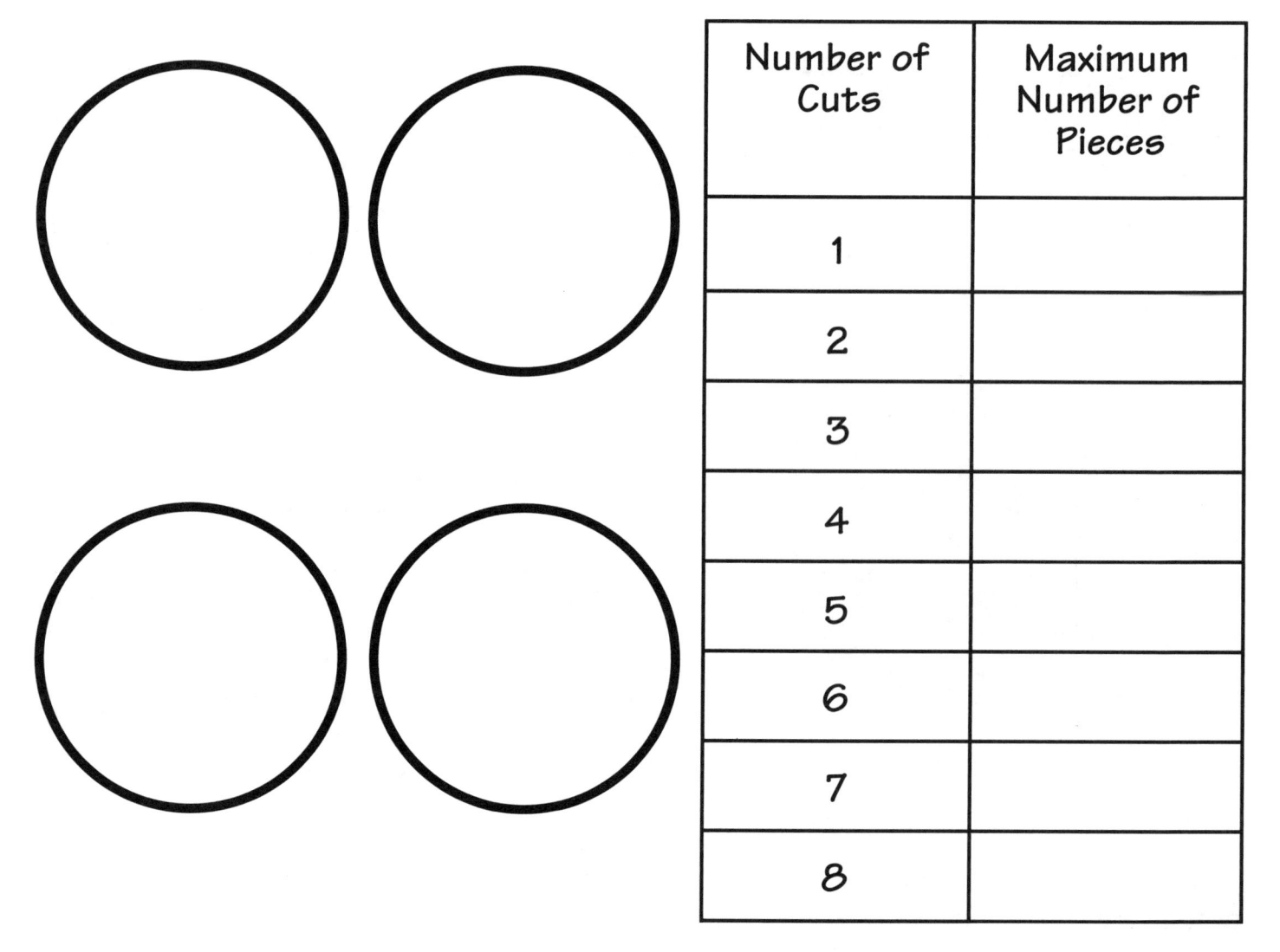

Number of Cuts	Maximum Number of Pieces
1	
2	
3	
4	
5	
6	
7	
8	

TOURNAMENT TIME

Find the number of games played in a single-elimination tournament with ten teams. Start by solving the problem for tournaments with 2 teams, 3 teams, and so on.

Discover a pattern in the relationship between the number of teams and the number of games, and use this to complete the table.

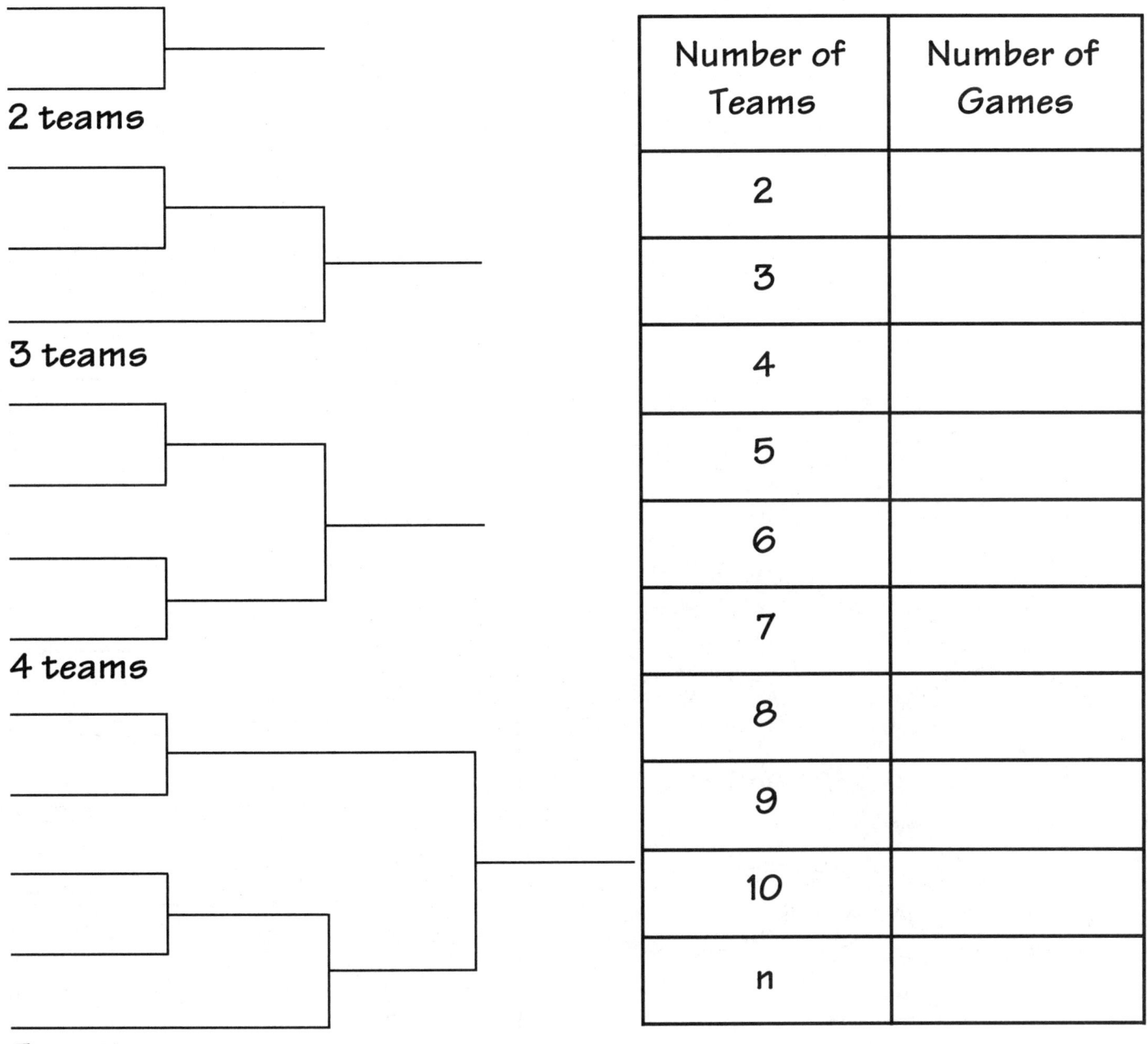

2 teams

3 teams

4 teams

5 teams

Number of Teams	Number of Games
2	
3	
4	
5	
6	
7	
8	
9	
10	
n	

HISTORICAL CONNECTIONS VOL. II   © 2005 AIMS Education Foundation

REGION REVENGE

If 10 points were placed on a circle so that the segments determined by these points created the maximum number of regions, how many regions would be created?

Count the regions in a circle with 1 point, 2 points, 3 points, and so on until you see the pattern.

Predict how many regions would be created in circles with 8, 9, and 10 points without actually drawing them.

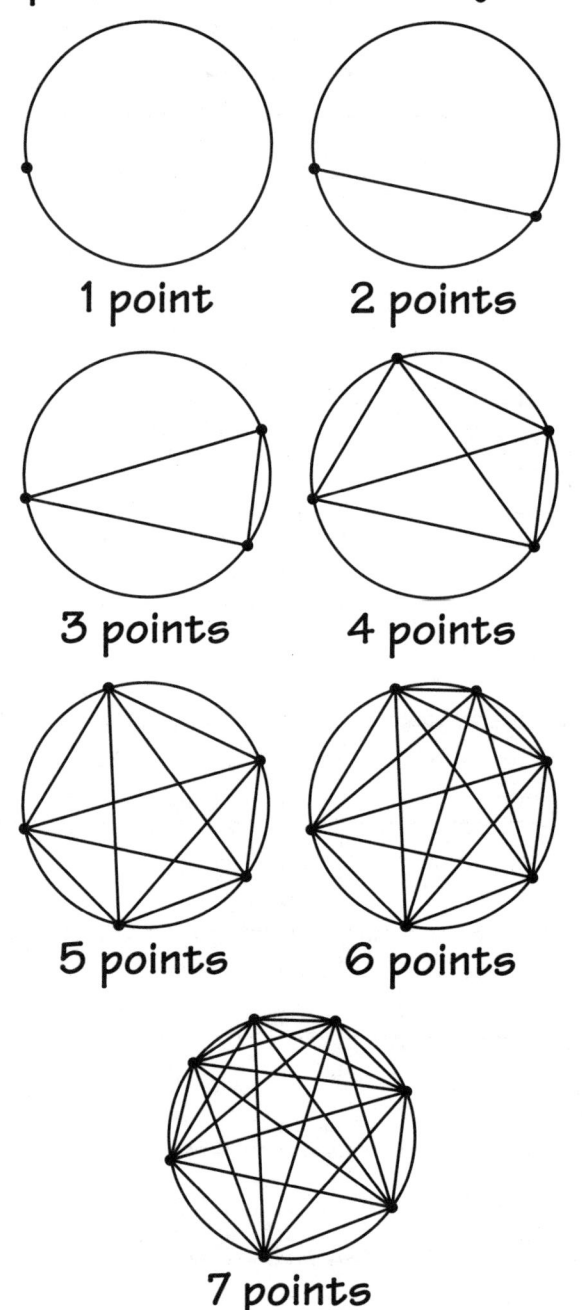

Number of Points	Maximum Number of Regions
1	1
2	2
3	4
4	
5	
6	
7	
8	
9	
10	

HISTORICAL CONNECTIONS VOL. II

CHECKERBOARD CHALLENGE

How many squares are there on an 8 by 8 checkerboard? Did you guess 64? Think again! How many squares of all sizes are there?

To solve this problem, count the squares on several smaller boards and record the numbers in the table.

Don't count squares on the larger boards! When you see the pattern, use it as a shortcut to complete the table.

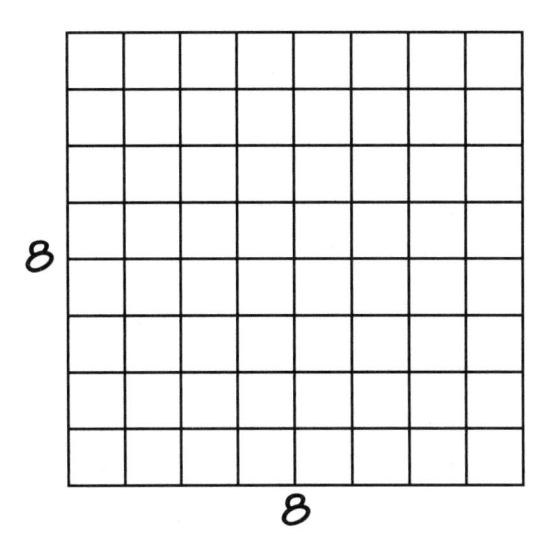

1 by 1

2 by 2

3 by 3

Dimension of Board	Number of Squares
1 by 1	1
2 by 2	5
3 by 3	
4 by 4	
5 by 5	
6 by 6	
7 by 7	
8 by 8	

Charles Babbage (1792-1871) designed an incredible "difference" machine that could continue patterns indefinitely. This machine is considered the first computer.

Galois
1811 - 1832

GALOIS
THE MISFORTUNATE MATHEMATICIAN

Biographical Facts:

Evariste Galois (gahl-WAH) was a French mathematician who contributed greatly to the understanding of algebra, laying the foundations of group theory. Galois was born on October 25, 1811, just outside of Paris in the village of Bourg-la-Reine. His father was both mayor and headmaster of the local boarding school. Galois' parents were well-educated, socially concerned persons; from them he learned to fight tyranny and injustice wherever it appeared. A series of unfortunate events kept Galois from receiving the education and recognition he deserved during his short life. Much of his most important work was recorded during the night before he was shot and killed in a duel on May 30, 1832. The circumstances of his death are still shrouded in intrigue, but the significance of his contribution to mathematics is very clear.

Contributions:

Galois' chief achievement was the solution to a problem that had puzzled mathematicians for centuries: under what conditions can an equation be solved? He proved that it is not possible to solve a general fifth (or higher) degree polynomial equation by radicals. His work was a remarkable breakthrough in the understanding of fields of algebraic numbers and groups. Today, he is universally considered the founder of modern group theory.

Anecdotes:

Problems in School

Galois' mother taught him at home until he was 12, when he was enrolled at the Louis-le-Grand school in Paris. This boarding school was a dismal place. With its barred windows and tyrannical supervisors, it felt more like prison than school. Here, he became rebellious and disinterested in learning. Because he refused to conform to the standardized regulations and expectations of the school, his teachers questioned his intelligence. They found him "strange" and "argumentative." Consequently, he was demoted and frequently disciplined, leaving him bitter and unhappy.

The Teacher Who Made a Difference

When Galois was 17, he met Louis-Paul-Emile Richard, a teacher of advanced mathematics at Louis-le-Grand. Richard was a great teacher, eager to do whatever he could to help students succeed. He quickly recognized Galois' genius and encouraged him to submit his work for review by Cauchy, one of the greatest mathematicians in France. Galois eagerly put together a paper that explained his fundamental discoveries in algebraic problem solving. Cauchy, who had agreed to present the work to the prestigious Academy of Sciences, unfortunately forgot about his promise and eventually lost the paper.

Entrance Examinations

Galois sought admission to École Polytechnique, the most highly respected math and science institute in all of Europe. He was very gifted at doing mathematics in his head, but the examiners wanted to see all the details of the problem-solving process on paper. The test results came back: "Admission denied." Some time later, he took the entrance examinations again. While some of the test administrators were impressed with Galois' capabilities, they had heard that he had an attitude problem. During the oral part of the exam, they began to test his humility by scoffing at his work and ridiculing his answers. In a burst of frustration, Galois grabbed a chalk eraser and hurled it at one of the examiners, hitting him in the forehead. It was an unpleasant end to Galois' chances to attend the best mathematics institute in Europe.

The Lost Grand Prize

At 19, Galois polished a collection of his work on the theory of algebraic equations and submitted it to the Academy of Sciences for the annual "Grand Prize in Mathematics" award. The project was highly original and significant, surely worthy of the prize. The Secretary of the Academy took Galois' entry home with him for evaluation, but the man unexpectedly died. When others went to his home in search of Galois' papers, they were not able to find them. Another opportunity to establish himself in the mathematical world was lost.

Political Poems and Ruinous Rhymes

Galois' father was a cultured, intellectual man who shared with his son the joy of making up lightly satirical verses for special occasions. Both father and son often delighted party-goers with their witty lines. Sometimes they used them to make a political point. One young priest, offended by the elder Galois, determined to get even. He composed a set of openly derogatory verses, signed the elder Galois' name to them, and circulated them throughout the village. Galois' father, convinced that his reputation was ruined, slipped into Paris and committed suicide.

A Treacherous Toast

Discouraged and frustrated by his inability to succeed within the educational system, Galois began to fight the injustice in society. He joined the National Guard, and quickly made a name for himself as an outspoken proponent of the revolution. At a banquet of the Young Republicans, a liberal group fighting against oppression by the royalty, Galois stood to make a toast "to Louis Philippe." Everyone present knew that the toast was ironic, because the king's health was not Galois' wish. But spies in the crowd noted the knife in Galois' left hand, and interpreted his toast as a death threat. Galois was arrested the next morning. For lack of evidence, the court was forced to acquit him (the knife was just for cutting his chicken), but Galois' name was placed on the list of "dangerous radicals."

A Peculiar Prisoner

Several months after his first arrest, Galois was imprisoned again. The charge? He had been caught wearing his National Guard uniform after his unit had been disbanded. Galois used six months in jail to work on his mathematical discoveries. His cell mates teased him and jeered because he refused to drink or play cards with them. He ignored them, preferring to keep his mind clear for his work.

The Deadly Duel

On the day Galois was released from prison, two thugs he had never seen before accosted him and challenged him to a duel to be fought the next morning at dawn. One of the men claimed that Galois had been flirting with his girlfriend. Galois was not interested in the girl, and did not want to fight, but honor required that he accept the challenge. Later evidence suggested political motives behind the duel.

Galois rushed to his room and began to write feverishly, somehow sensing that this night was his last. He wrote a letter to his friend, Chevalier, in which he recorded many of his mathematical discoveries. Galois urged Chevalier to share these ideas with Jacobi and Gauss, two eminent mathematicians. All night long he worked, trying to preserve on paper some of the many ideas in his mind. In the margins next to a theorem, he scrawled, "There are a few things left to be completed in this proof. I do not have time." Finally, after explaining the situation and bemoaning the trivial circumstances calling for his death, he closed the letter with a prayer: "Eventually there will be, I hope, some people who will find it profitable to decipher this mess."

At dawn, Galois went alone to the appointed place. The referee counted as the two young men marched 20 paces. Then he shouted "Ready, aim!" and Galois fell to the street, shot through the abdomen. He lay there until 9:00 a.m. when a passerby found him and took him to the hospital. His wound became infected. Galois died the next day and was buried in a common ditch in the Montparnasse cemetery. He was 20 years old.

ROTATING TIRES: A MATHEMATICAL EVENT!

Galois (1811-1832), a French mathematician, made major contributions to several branches of mathematics including group theory. Galois' work gave mathematicians new tools for classifying mathematical systems.

Let's think about four possible ways to rotate the tires on a car. The four rotations along with the table is an example of a group.

ROTATION 0   ROTATION 1   ROTATION 2   ROTATION 3

Suppose we perform Rotation 1 followed by Rotation 3. The end result is the same as if we had done only Rotation 2. We symbolize this as 1*3 = 2. To see that this is true, cut out the four tires at the bottom of this sheet and move them, performing Rotation 1 followed by Rotation 3.

Use the four tires to help you complete the table on the right. Perform the rotation in the left column first. Then do the one in the top row.

Record in the table the single rotation that accomplishes the same thing as these two rotations.

LF: Left Front   RF: Right Front
LR: Left Rear   RR: Right Rear

*	0	1	2	3
0				
1				2
2				
3				

HISTORICAL CONNECTIONS VOL. II   © 2005 AIMS Education Foundation

CLOCK ARITHMETIC

Under what circumstances would 9 + 4 = 1?
Could 8 + 5 = 1 also?

When determining time, addition follows a new set of rules. Use the clock above to complete this addition table.

+	1	2	3	4	5	6	7	8	9	10	11	12
1	2	3	4	5	6	7	8	9	9	11	12	1
2	3	4										
3	4		6									
4	5			8								
5	6				10							
6	7					12						
7	8						2					
8	9							4				
9	10								6			
10	11									8		
11	12										10	
12	1											12

1. Dr. Smith prescribed some penicillin for Josh. He is supposed to take a dose every six hours. If he first takes it at 7 a.m., when should he take the next capsule?
2. Jennifer is flying north to her grandmother's home for the holidays. Her plane leaves at 9 a.m., and the flight takes 7 hours. If she stays in the same time zone, what time should she arrive?

Clock arithmetic is an example of what is called a group in mathematics. Groups were studied by Galois and play a very important role in mathematics.

MAGIC FLEXAGONS

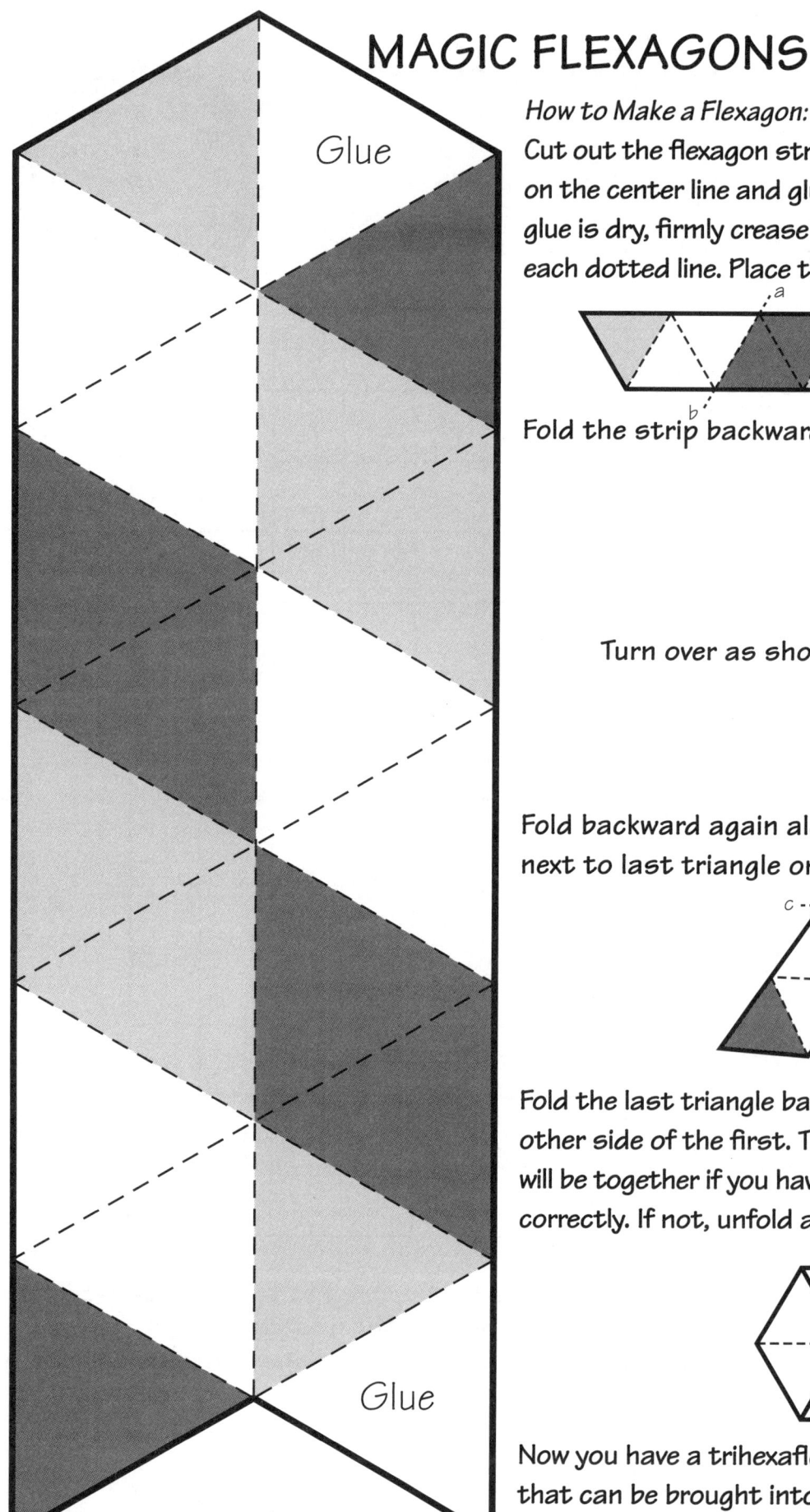

How to Make a Flexagon:

Cut out the flexagon strip on the left. Fold carefully on the center line and glue with rubber cement. When glue is dry, firmly crease the strip back and forth on each dotted line. Place the strip as shown.

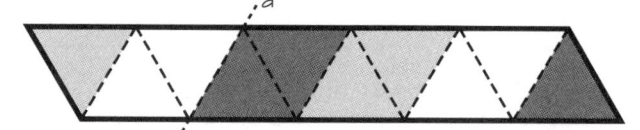

Fold the strip backward along the line ab.

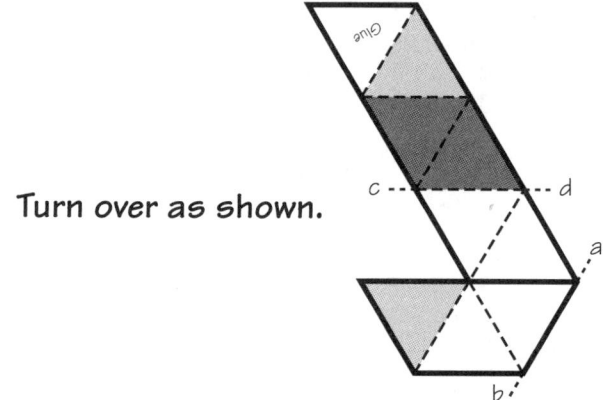

Turn over as shown.

Fold backward again along the line cd, placing the next to last triangle on top of the first.

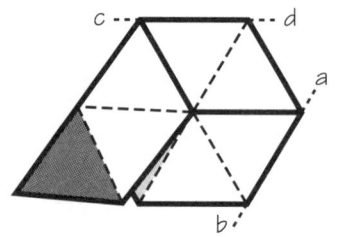

Fold the last triangle backward and glue it to the other side of the first. The triangles marked "glue" will be together if you have constructed the flexagon correctly. If not, unfold and start again.

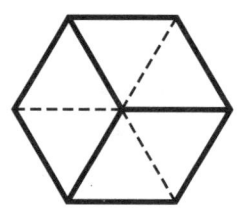

Now you have a trihexaflexagon: "tri" for three faces that can be brought into view, "hexa" for its hexagonal form, and "flexagon" for its ability to flex.

HISTORICAL CONNECTIONS VOL. II © 2005 AIMS Education Foundation

MAGIC FLEXAGONS

How to Flex a Flexagon:

To flex down, start with the flexagon in an open position. Bring three alternate vertices together at the base of the flexagon. Gently open the flexagon at the top, revealing a new color. If it doesn't open easily, try the other three vertices. →

← To flex up, start with the flexagon in an open position. Bring three alternate vertices together at the top of the flexagon. Use your fingers to gently open the flexagon from the bottom.

Note: Firmly creasing the folds will make the flexagon flex more easily.

Flexagons are lots of fun! They can also be used to illustrate what is called a group in mathematics. Here is an activity that demonstrates group properties.

To do this activity, always start with the flexagon in a horizontal position, with the gray face (G) on top and the black face (B) on the bottom. Perform the transformations indicated by the codes and note the resulting positions.

Code Box
N = do nothing, yields GB (gray on top, black on bottom)
↓ = flex down, yields WG (white on top, gray on bottom)
↑ = flex up, yields BW
◯ = turn over, yields BG
⬇̇ = flex down and turn over, yields GW
⬆̇ = flex up and turn over, yields WB

✱	N	↓	↑	◯	⬇̇	⬆̇
N						
↓						◯
↑						
◯						
⬇̇						↑
⬆̇			⬇̇			

Suppose we perform ↓ followed by ⬆̇. The result is BG, which according to the Code Box, is equal to ◯. We symbolize this as ↓ ✱ ⬆̇ = ◯. To see that this is true, perform the operation with your flexagon.

Complete the table using your flexagon. First perform the transformation in the vertical column, then the one in the corresponding horizontal column. Refer to the Code Box and record the symbol for the equivalent transformation. The six transformations in the Code Box along with the table form a group.

HISTORICAL CONNECTIONS VOL. II © 2005 AIMS Education Foundation

THE ERASED CHANCE
A Skit to Read

Narrator 1: Evariste Galois (ay-va-REEST gahl-WAH) was one of the brightest boys in France. He wanted to get into Ecole Polytechnique (ay-COLE polly-tek-NEEK). It was the best school in Europe, a place where he could develop his mathematical interest. But getting in wasn't easy.

Narrator 2: You see, Galois had this habit of doing math in his head. He didn't like to waste time showing his work. The process was obvious to him. If other people couldn't see it, that was their problem. He also got very impatient when teachers wanted him to read about ideas he already understood or solve problems he had already mastered.

Narrator 1: Everyone who wanted to go to Ecole Polytechnique had to take a difficult test. Galois went to take the exam and quickly answered the questions. It wasn't even hard for him. He finished early and looked around at the other applicants still struggling with the problems. He checked over his test paper to make sure everything was correct, and left.

Narrator 2: Galois waited a long time for his acceptance letter from the school. Finally it arrived. He tore open the envelope and gasped. There, in big letters under his name, were two horrible, shocking words: "ADMISSION DENIED."
Galois went to the school to see if there could possibly be an error. The chief examiner shook his head.

Examiner 1: "No. No error. Since you didn't show your work, we have every reason to believe you cheated. Perhaps you just copied the answers from your neighbor!"

Galois: "But sir, I do my work in my head. It's my pattern; I always do it that way. Won't you reconsider? I want so much to attend your school!

Examiner 1: "Sorry! You should have thought of that before! If you want to take the test again, you will have to wait until next year. But don't goof up next time. Remember, you only get two chances."

Narrator 1: Galois was deeply disappointed, but he determined to prepare and take the test again. The year passed slowly. Finally the day arrived. When the examiners heard that Galois would be re-taking the test, they got together and talked about him.

Examiner 1: "He's a real smart aleck."

Examiner 2: "That's for sure. I've heard that he boasts about knowing more than we do!"

Examiner 3: "Well, we'll show him!"

Narrator 2: Galois completed the written part of the test without any problems. He tried to show more of his work than he had the first time. Then he was called into a small room for the oral part of the examination.

Examiner 2: "Young man, what do you think of the way mathematics is being taught in the schools today?"

Narrator 1: Galois paused. Should he be honest? His parents had taught him that honesty was always the best policy.

Galois: "Well, sir, with all due respect, I think most teachers today kill any desire to study mathematics students might have. They use old books and boring techniques."

Examiner 3: "Humph! Is that so? That's quite an attitude for a boy your age."

Examiner 1: "So what have *you* found interesting in mathematics?"

Narrator 2: Galois was much happier with this question. He began to share what he had discovered about which equations could be solved using algebra and which could not. Soon he noticed that his examiners were winking at each other and making exaggerated faces. When he paused to ask for questions, the examiners began to laugh.

Examiner 2: "Who do you think you are, pretending to solve a problem which has stumped mathematicians for centuries?"

Examiner 3: "Yes. You're getting too big for your britches, boy!"

Examiner 1: "Let's move on, gentlemen. We have *serious* candidates to interview."

Narrator 1: Galois felt his face getting hot. His knees began to shake, and his breathing came hard. He got so frustrated when the examiners continued to taunt him, that he reacted without thinking. He grabbed a chalk eraser off the table in front of him, and hurled it at one of the laughing examiners. It hit its mark—smacking the man right on the forehead.

Narrator 2: But it was no bull's eye for Galois. Even though his math was right and his discoveries were remarkable, Galois' act erased any chances he might have had of getting into the school.

Lovelace
1815 - 1852

ADA BYRON LOVELACE
The First Computer Programmer

Biographical Facts:

Ada Byron Lovelace was a British mathematician and musician, born in London in 1815. Her father was the British poet, Lord Byron. Her mother, Annabella Milbanke, encouraged her to study mathematics. Ada married Lord William King, Earl of Lovelace, and had three children. She died of cancer in 1852 at the age of 36.

Contributions:

Ada Lovelace is best known as the first computer programmer. She wrote about Charles Babbage's "Analytical Engine" with such clarity and insight that her work became the premier text explaining the process now known as computer programming.

Anecdotes:

A Famous Father

Lord Byron was a flamboyantly handsome man who traveled widely and wrote what was sometimes biting criticism of British society. He fell in love with Annabella Milbanke, but their marriage lasted only one year. One week after his daughter Ada was born, Byron left for Italy and never saw Ada again. He died in Greece while fighting for Greek freedom from the Turks when Ada was 8 years old.

Those who knew Byron often remarked on the similarities between him and his mathematical daughter. The two shared dark, romantic good looks. They both died young, at exactly the same age—36. In very different ways, they both experienced periods of great achievement and accomplishment.

One of Byron's poems is about Ada. At her request, she is buried beside her father in Nottinghamshire.

Childhood Choices

Ada Lovelace was an active, energetic child. She loved gymnastics, dancing, and especially horseback riding. She became an accomplished musician, learning to play piano, violin, and harp. Her mother once commented that Ada was especially fascinated by mechanical things; she loved to figure out what made machines work.

As a teenager, Ada benefited from the advantages and activities of the upper class in London. She frequented concerts, theaters, and elegant parties. She met many famous and influential people, including the queen of England.

What's a Woman to Do?

The one person young Ada most longed to meet was Mary Sommerville, a mathematician who had just published *The Mechanism of the Heavens*, a book on mathematical astronomy. Fortunately, the two became friends. It was Mrs. Sommerville who arranged for Ada to meet Lord William King, who later became Ada's husband. For Ada, Mrs. Sommerville was a role model—a woman who was also a mathematician!

The Inspirational Engine

At 18, Ada Lovelace met Charles Babbage, who invited her to study his difference engine. By observing what Babbage had designed and by asking him questions, she soon became an expert on the inventor's work. When Babbage changed his plans and began to design his analytical engine, Lovelace saw tremendous potential in the machine. She understood it better than most people older and more experienced than she.

The Italian mathematician Menabrea had attempted to explain how the analytical engine would work in a presentation at a scientific conference in Vienna. Lovelace was asked to translate his paper into English. While doing so, she added footnotes and explanatory sections that greatly enhanced the original. By the time she was finished, the paper was three times as long as Menabrea's, and much more useful.

Babbage was very pleased. He published and distributed Lovelace's work, modestly signed with only her initials, "A.A.L." Although this paper was the summit of her career, she felt it was unbecoming for a woman of her social class to publish anything so "unfeminine." It was nearly 30 years before the identity of "A.A.L." was commonly known.

Could Computers Compose?

Throughout her career as a mathematician, Lovelace was often drawn by her love of music. She had shown much promise as a young woman and many of her friends expected her to pursue musical studies. As she explored the powers of the analytical engine, she thought about the potential of such machines to compose music. Now, a century after she predicted it, her dream has come true. Computers are composing music!

Off to the Races

Charles Babbage and Ada Lovelace had another common interest besides computers. Together, they had come up with what they considered an "infallible system" for beating the odds at the horse races. Babbage needed money to fund the construction of his engine; Lovelace was simply a compulsive gambler. Their system did not work, and both of them were disgraced by massive gambling debts and associations with "shady" bookmakers.

ADA: A New Language

The computer language, ADA, was commissioned in the late 1970s by the United States Department of Defense. Based on the language PASCAL, ADA is a general-purpose language designed to be readable and easily maintained. It is efficient for machines, yet easy to use. It was intended to become a standard language to replace the many specialized computer languages now in use.

CALCULATOR FUN

Perform each of the indicated computations on a calculator. Then turn the calculator upside down and read the word answer. A clue is given for each problem.

Calculation	Numerical Answer	Clue	Word Answer
1. (354 x 15) + 7	_____	What you should never tell	_____
2. 48,450 ÷ 6	_____	A messy person	_____
3. $88^2 - 3^2$	_____	Opposite of buy	_____
4. 463 x 79 - 1469	_____	The capitol of Idaho	_____
5. 1911 x 3	_____	Snakelike fish	_____
6. $19^3 + 879$	_____	It rings	_____
7. 106 x 35 - 5	_____	The bottom of a shoe	_____
8. $84^2 + 7^2$	_____	To make dirty	_____
9. 1377 x 4	_____	Person in charge	_____
10. 2875 + 2463	_____	They sting	_____
11. 250 x 140 + 7	_____	Opposite of tight	_____
12. 25 x 2564 - 6382	_____	Have to be paid each month	_____

Make up some problems like this of your own.

Calculators and computers have made a dramatic contribution to mathematics. Blaise Pascal invented the first mechanical calculator in 1642. Ada Lovelace (1815-1852) was the first computer programmer. Two of our computer languages, PASCAL and ADA, are named after them.

CIRCLE CIRCUS

Find the maximum number of interior regions created by intersecting circles.

Carefully count the regions in the circles below until you see the pattern. Complete the table by predicting (not drawing) how many regions will be created by 6, 7, and 8 intersecting circles.

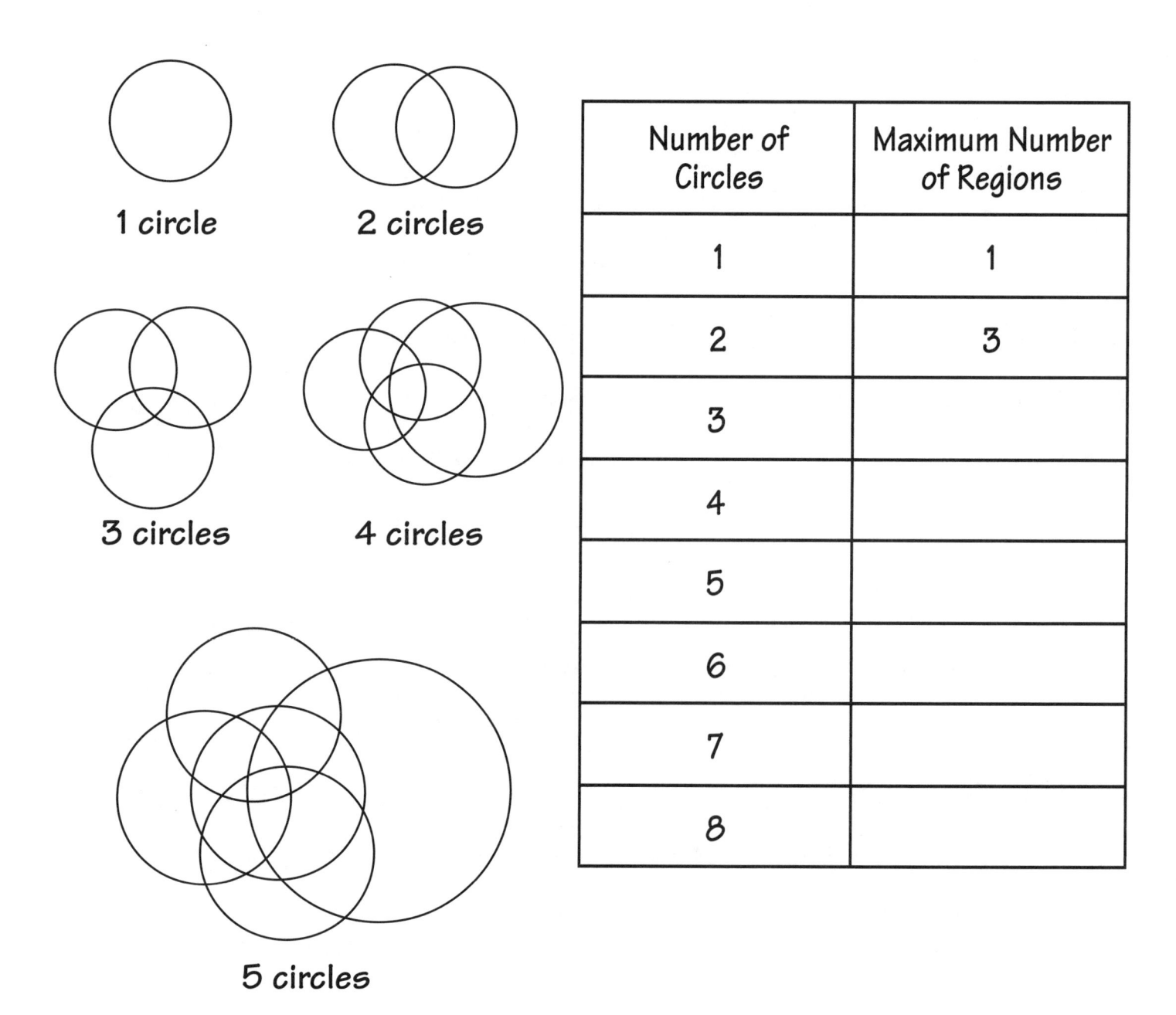

1 circle

2 circles

3 circles

4 circles

5 circles

Number of Circles	Maximum Number of Regions
1	1
2	3
3	
4	
5	
6	
7	
8	

Can you find a formula which will give the number of regions for n circles?

TWENTY-ONE CONNECT

Here is a circle with 21 points joined by all possible line segments. Find the total number of line segments in the drawing by first solving several easier problems.

Count the number of line segments in circles with 2 points, 3 points, and so on, and record your findings in the table.

Discover the pattern and solve the problem.

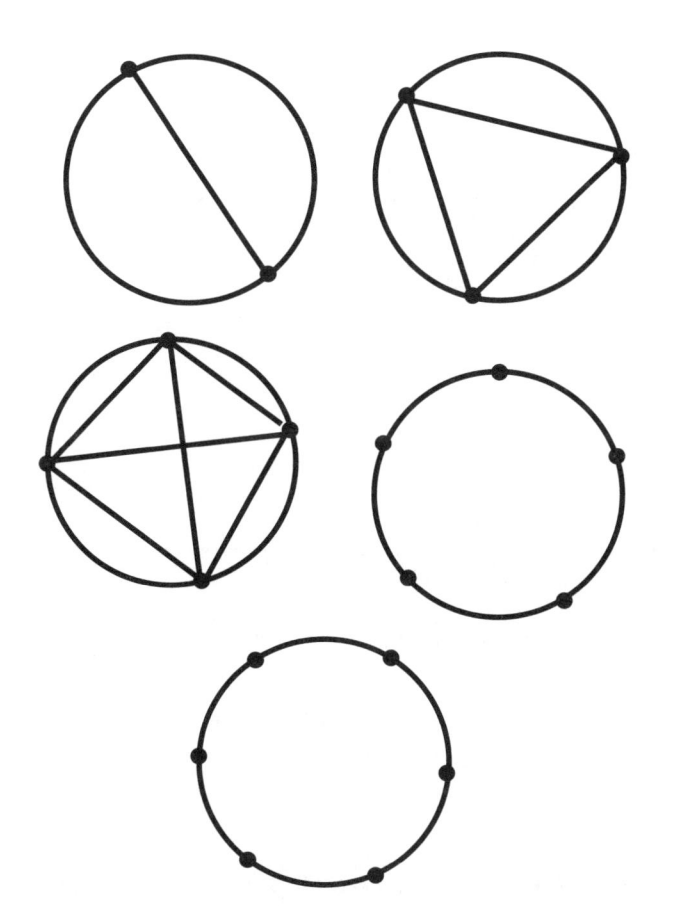

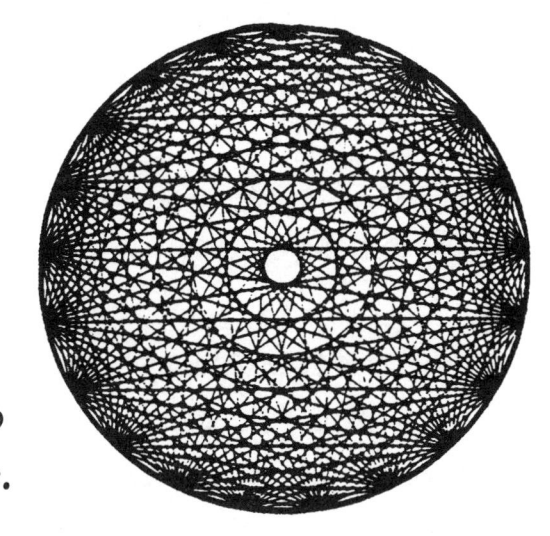

Number of Points	Number of Segments
2	1
3	3
4	
5	
6	
7	
21	
n	

Can you discover a formula for a circle with n points?

A beautiful design can be made by winding string or yarn in every possible way around 21 nails evenly spaced on a circular board.

HISTORICAL CONNECTIONS VOL. II © 2005 AIMS Education Foundation

LEAP FROG

Pairs of pegs (white and black) are placed on a peg board, with one extra hole in the middle of the board.

The goal of this activity is to interchange the positions of the black pegs and the white pegs according to the following rules:

(a) white pegs always move right, black pegs left,
(b) only one peg can be moved at a time,
(c) a peg can move to an adjacent hole, or
(d) a peg can jump over one peg of the opposite color to an empty hole.

Record the number of moves required on peg boards of various sizes.

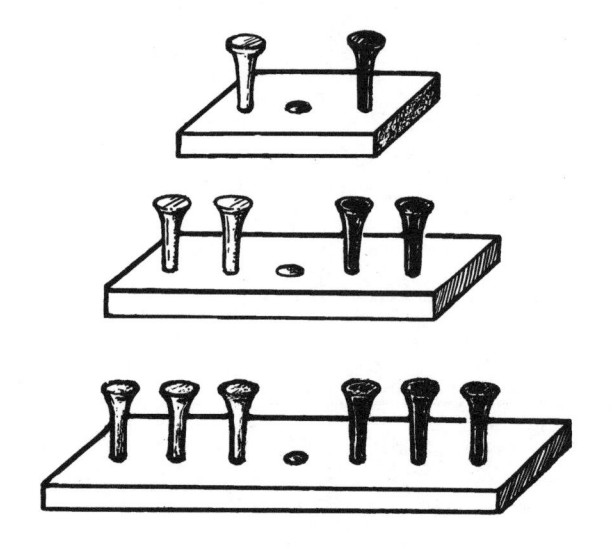

Number of Pegs on One Half	Minimum Number of Moves
1	
2	
3	
4	
5	
20	
n	

Discover a pattern to complete the table.

If a peg board is not available, the problem can also be solved using pennies and dimes or colored chips. Place them on the provided diagram.

HISTORICAL CONNECTIONS VOL. II   © 2005 AIMS Education Foundation

ROUND ROBIN PLAY

Each team in a volleyball league plays each of the other teams once.

Suppose the league is made up of eight teams.

What is the *total* number of games played?

Solve this problem by first looking at some easier problems.

Count the number of games required for 2 teams, 3 teams, and so on, and observe the pattern.

Can you find the formula for n teams?

Number of Teams	Number of Games
2	1
3	
4	
5	
6	
7	
8	
n	

Kovalevsky
1850 - 1891

SONYA KOVALEVSKY
DARING AND DETERMINED

Biographical Facts:
Sonya Corvin-Krukovsky Kovalevsky (kah-vuh-LEV-ski) was born on January 15, 1850, in Moscow. Her father was a general in the Russian army, and both of her parents were well-educated. Sonya received most of her training at home until, as a young adult, she studied in several universities throughout Europe. She had one child, Fufa, born in October, 1878. Kovalevsky died quite suddenly of pneumonia in 1891, at the age of 41. She is buried in Stockholm, the only city in which her professional life was affirmed.

Contributions:
Kovalevsky has been called "the most dazzling mathematical genius to surface among women during the past two centuries" (Lynn Osen). She was extraordinarily versatile and talented, and penetrated deeply into methods of mathematical research. While Kovalevsky's central concern was with infinite series, she contributed significantly to the understanding of Abel's functions and partial differential equations. Her work on Saturn's rings greatly advanced understanding of that planet.

Anecdotes:
The Lonely Child
When Sonya was about six years old, her father retired from the army and the family moved to Palibino, a large estate near the Lithuanian border. Living in an ancient feudal castle in such a remote location accentuated Sonya's feelings of being unloved and ignored by her parents. She and her brother and sister were each given a governess or tutor, and the children were kept separated most of the time.

From age 5 to 12, Sonya was under the guidance of a strict English governess. She made Sonya take long daily walks in the winter unless the temperature was colder than 10 degrees. Sometimes they heard the chilling howls of wolves. If Sonya disobeyed or did not perform up to expectations, her teacher humiliated her by pinning insulting signs on the back of her clothes before dinner.

The Writing on the Wall
The Krukovskys ordered elegant wallpaper for their spacious country home but, unfortunately, not enough to cover all the walls. Consequently, Sonya's room was papered with lithographed differential and integral calculus lecture notes. Her father had purchased these

during his student days, and he thought they should be put to use. Little did the family realize what an effect this would have on Sonya. She spent hours trying to decipher the formulas and text, analyzing the squiggles and mentally rearranging the pages into their correct order. Later, when she was about 15, she studied differential calculus for the first time. Her teachers were amazed at how quickly she mastered the subject, almost as if she had seen it before!

An Unusual Proposal

Russian universities were closed to women. Those who wanted to continue their educations had no choice but to study in foreign countries. Respectable Russian families, however, would never let their single daughters travel abroad without escort. Sonya and her sister, Aniuta, concocted a plan used by hundreds of young Russian girls: they would find a young male student to marry and use him as a "passport."

Since Aniuta was older than Sonya, she was to marry Vladimir Kovalevsky, a promising geology student who planned to study in Germany or Switzerland. Sonya would be allowed to accompany them as a sister. The young man agreed to the arrangement on the condition that he marry Sonya instead; he was intrigued by her expressive eyes and quick wit.

Papa Says No

General Krukovsky refused to allow Sonya to marry. After all, Aniuta was six years older and still single. It wouldn't be proper! But the sisters were determined. One night Sonya ran away from home to Vladimir's apartments, leaving a note for her father. To save his family from disgrace, the general was forced to announce their engagement and six months later they were married.

Learning with the Masters

Although Sonya and Vladimir did not live together as husband and wife for five years, they studied at universities in European cities including Vienna, Heidelberg, Paris, and Berlin.

In Heidelberg, Vladimir studied paleontology, and Sonya attended math and physics lectures. She learned of the work of R.W. Bunsen, the inventor of the gas burner, and wanted to study with him. But Bunsen had sworn never to let a female into his lab. Sonya had a remarkable capacity to get such people to change their minds, and the two became friends and co-workers.

She was most profoundly influenced by her work with Karl Weierstrass, the "father of mathematical analysis." Since the university at Berlin would not admit women, Sonya asked Weierstrass to take her on as a private student. He was shocked by her audacity and gave her a set of difficult problems, probably to get rid of her. Her solutions were not only correct but remarkable, and he eventually became one of her strongest supporters.

The University of Göttingen granted Sonya Kovalevsky the doctoral degree *summa cum laude* in 1874.

Stockholm and Strident Strindberg

In spite of her growing reputation, it took much persuasion and persistence for Sonya Kovalevsky to receive an appointment to a major university. But when she arrived to begin teaching at the University of Stockholm, many parties and balls were given in her honor. She was the "Princess of Science," the only woman professor in the country. Hundreds of baby girls were named "Sonya" in her honor.

One of the few persons who did not appreciate Sonya Kovalevsky was August Strindberg, the famous Swedish playwright. Threatened, perhaps, by her feminist perspectives, Strindberg wrote, "She proves, as decidedly as that two and two make four, what a monstrosity a woman professor of mathematics is, and how unnecessary, injurious and out of place she is."

Time Out for Literature

In addition to her achievements in math and science, Sonya Kovalevsky was also a talented writer. She wrote short stories, magazine articles, and poetry in at least three languages: French, Swedish, and Russian. Much of what we know about her early life comes from her autobiography, *Recollections of Childhood*. She wrote several novels including *The Rayevsky Sisters,* published in 1889, and *Vera Votontzoff.*

During her travels, Kovalevsky had many opportunities to meet influential persons from the literary world. She became good friends with George Eliot, and wrote a brief biography of her in 1886. She also spent time with Charles Darwin, Thomas Huxley, and Fyodor Dostoevsky.

The Prix Bordin
In 1888, Kovalevsky entered the Paris Academy of Sciences' most prestigious competition, the "Prix Bordin." She spent much of the summer working on her project on Saturn's rings. The contest was conducted with integrity; none of the judges knew who had submitted the entries. Each contestant was instructed to seal his or her name inside an envelope, and then to write a motto on the sealed envelope with a corresponding motto on the document itself.

When the judges reached a decision, they announced that the winning entry was so exceptional that the usual prize of 3000 francs had been increased to 5000. The envelope was opened on Christmas Eve, and the grand prize winner was Sonya Kovalevsky!

Her motto:
"Say what you know,
Do what you must,
Come what may"

Leaving Her Mark
Sonya Kovalevsky is the only Russian woman in mathematics to have a postage stamp issued in her honor. There is also a landmark on the moon named "Kovalevsky."

THE SNOWFLAKE CURVE

Would you believe that a geometric figure with area less than 1 square inch can have a perimeter of more than 100 million miles? The "snowflake" is an example of such a curve.

Here's how to make one: Start with a 1-inch equilateral triangle. Divide each side of this triangle into thirds. On each middle third, build another equilateral triangle. Continue this process indefinitely. The curve soon begins to resemble a snowflake!

The first four stages of this sequence are shown.

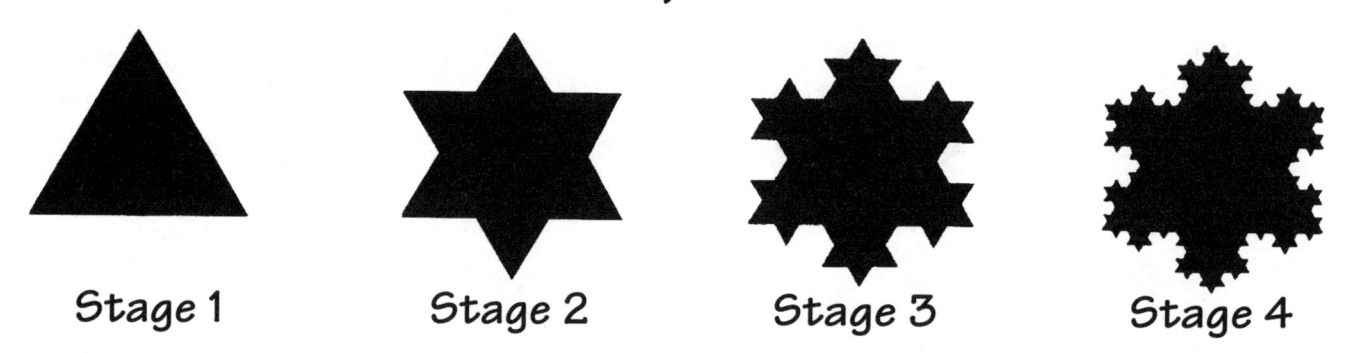

Stage 1   Stage 2   Stage 3   Stage 4

To find out what happens to the perimeter of the curve, complete the table below. Assume each side of the original equilateral triangle to be 1 inch.

Careful observation of the relationships in the table will help you discover the formulas for the nth stage.

Stage Number	Number of Sides	Length of each side (in inches)	Perimeter of Snowflake (in inches)
1	3	1	$3 \cdot 1 = 3$
2	12	$\frac{1}{3}$	$12 \cdot \frac{1}{3} = 4$
3	48	$\frac{1}{9}$	$48 \cdot \frac{1}{9} = 5\frac{1}{3}$
4	192	$\frac{1}{27}$	$192 \cdot \frac{1}{27} = 7\frac{1}{9}$
5			
n			

Use a scientific calculator to find the perimeter (in miles) of the snowflake created by the (a) 60th stage and (b) 100th stage.

HISTORICAL CONNECTIONS VOL. II © 2005 AIMS Education Foundation

PREDICTABLE PATTERNS

Sonya Kovalevsky (1850-1891) was fascinated by infinite sequences. Follow her example by filling in the spaces to continue these sequences.

1. 1, 3, 5, 7, 9, _____, _____, _____, _____

2. 4, 8, 12, 16, 20, _____, _____, _____

3. 1, 2, 4, 7, 11, _____, _____, _____, _____

4. 1, 3, 6, 10, 15, _____, _____, _____

5. 1, 4, 9, 16, 25, _____, _____, _____, _____

6. 1, 2, 4, 8, 16, _____, _____, _____, _____

7. 3, 6, 9, 12, _____, _____, _____, _____

8. 3, 2, 6, 5, 15, 14, _____, _____, _____, _____

9. 1, 3, 2, 4, 3, _____, _____, _____, _____

10. 2, 3, 7, 16, 32, _____, _____

11. 1, 1, 2, 3, 5, 8, 13, 21, _____, _____, _____, _____

12. 3, 5, 9, 15, 23, 33, _____, _____, _____

13. 1, 2, 5, 10, 17, 26, _____, _____, _____

14. 1, 4, 9, 1, 6, 2, 5, 3, 6, 4, 9, 6, 4, _____, _____, _____, _____

15. 1, 1, 4, 3, 9, 6, 16, 10, 25, 15, _____, _____, _____, _____

HISTORICAL CONNECTIONS VOL. II   73   © 2005 AIMS Education Foundation

THE CHESSBOARD COVERED WITH WHEAT

According to ancient legend, the king of India wanted to show appreciation to the servant who had invented the game of chess for the king's entertainment.

"All I ask, your majesty," said the servant, "is some grains of wheat for each square on the chessboard. Give me one grain on the first square, two on the second square, four on the third, and so on. Double the number from one square to the next until all 64 are covered."

The request seemed simple enough, and the king quickly agreed.

Complete the table below to find out what kind of trouble the king ran into.

Number of Squares	Total Number of Grains
1	1
2	3
3	
4	
5	
6	
64	

Amazing Fact!
To fulfill his promise, the king would have needed more wheat than the world has produced in the last 2000 years. This much wheat, spread over the entire Earth, would be $1\frac{1}{2}$ inches deep.

PAPER PUNCHING PATTERNS

Here is an interesting activity using a sheet of paper and a hole punch.

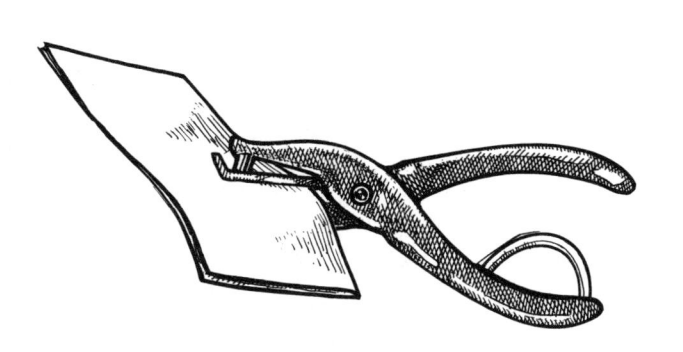

1. Punch one hole in a sheet of paper.

2. Fold the paper in half and punch through it in a new spot.

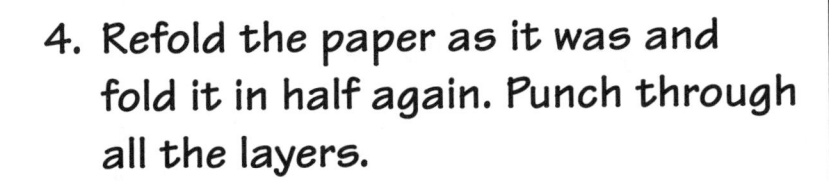

3. Unfold the paper and record the total number of holes.

4. Refold the paper as it was and fold it in half again. Punch through all the layers.

5. Guess the total number of holes. Unfold the paper, check your guess, and record the result in the table.

6. Continue folding, punching, and counting until the folded paper becomes too thick.

7. Use the pattern in the second column to predict the number of holes for the rest of the table.

Can you write a formula for n folds?

Number of times folded in half	Number of holes
0	
1	
2	
3	
4	
5	
6	
7	
8	
20	
n	

HISTORICAL CONNECTIONS VOL. II © 2005 AIMS Education Foundation

Ramanujan
1887 - 1920

RAMANUJAN
INDIA'S MATHEMATICAL GENIUS

Biographical Facts:

Srinivasa Ramanujan (rah-MAH-nuh-jun) was an Indian mathematician who, without help or formal education, made some of the most amazing discoveries in the history of mathematics. He was born in 1887 to a poor family in southern India.

Because he focused so exclusively on mathematics, Ramanujan was unable to pass other college courses, and his formal education ended early. He was "discovered" by the British mathematician G. H. Hardy, and worked in England with the leading scholars of the time. Unable to adjust to British culture and lifestyle, he became ill and returned to India where he died in 1920 at the age of 32.

Contributions:

Ramanujan had an uncanny ability to see through intricate number relations and reduce the complex to the simple. He discovered formulas for finding π as well as many other theorems and formulas. Much of his work is relevant to problems he could have known nothing about. Recently, Ramanujan's work has been found to have application to string theory in physics and to fast algorithms in computer science. The more his work is studied, the more remarkable it becomes.

$$\frac{1}{\pi} = \frac{\sqrt{8}}{9801} \sum_{n=0}^{\infty} \frac{(4n)!\,(1103 + 26390n)}{(n!)^4 \, 396^{4n}}$$

One of Ramanujan's Remarkable Formulas for π

Anecdotes:

Humble Beginnings

Many factors were against Ramanujan in childhood: he was shy and found it difficult to speak, he suffered through a severe case of smallpox, and his small home was always crowded with the boarders his mother took in to help cover expenses. Nevertheless, he did well in school, revealing his mathematical genius by reciting formulas to his classmates.

At seven, he was given a scholarship to high school. When the annual award in mathematics was presented, it always went to Ramanujan. Once, the prize was a book of Wordsworth poetry. Poetry was okay, but he would have much preferred a book about mathematics; he read his first one at the age of 12!

Education? Two Books!

Ramanujan's entire higher mathematics education came from two borrowed books: S.L. Loney's *Plane Trigonometry* and G.S. Carr's *Synopsis of Elementary Results in Pure Mathematics*. The latter was an unusual text in that it contained only the results of as many as 6000 theorems. Carr used it when he worked as a tutor at Cambridge, but without him there to explain the problem-solving process, it would have been useless to most readers.

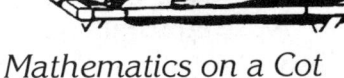

Mathematics on a Cot

Ramanujan had some rather unusual work habits. He liked to work problems with chalk on a slate, erasing his calculations with his elbow as he went. Much to the chagrin of mathematicians today, he did not show his work. Everything was erased except the final formulas.

Ramanujan's favorite position for doing mathematics was lying on his stomach on a cot set up on the veranda. He would scribble page after page of equations, refusing to stop even to eat or sleep. Sometimes his wife or his mother would come and feed him while he worked so he would not have to interrupt his thoughts.

Is This a Crackpot?

Ramanujan worked as a clerk by day and pursued mathematics at night. His employers urged him to send his mathematical discoveries to England for evaluation. Two of the three mathematicians he corresponded with returned his letters without comment, apparently writing him off as a crackpot. His work was like nothing they had ever seen. It appeared disorganized and chaotic, as if it were based on intuition rather than mathematical proof.

The third mathematician Ramanujan wrote to was G. H. Hardy. At first he was also skeptical, but he soon became intrigued with this unknown scholar's remarkable yet unorthodox formulas. Ramanujan's ideas must be true, said Hardy, because "no one would have had the imagination to invent them." Hardy invited Ramanujan to England so that Hardy could meet and collaborate with a man he considered not a crackpot, but a genius.

Travel Plans Revealed in Dreams

Hardy did not have an easy time persuading Ramanujan to come to England to work and study with other mathematicians. First of all, it was against his Brahmin caste which taught that overseas travel would defile its members. Such persons would not be allowed to attend religious ceremonies such as weddings or funerals. Because of this, Ramanujan's strong-willed mother was adamantly opposed to his going.

The son was submissive and agreed to stay home. But then something strange happened: Ramanujan's mother had a dream in which the family goddess proclaimed Ramanujan should go abroad. He packed his trunks and left family and friends, boarding a ship for England.

The Ramanujan-Hardy Connection

Ramanujan arrived at Trinity College, Cambridge, in April, 1914. He tried hard to adjust to European ways, cutting his hair and trading his turban for a hat. But he was lonely, having left his wife and everything familiar behind. He cooked his own meals in his room, but had trouble finding enough fresh vegetables and fruits for his vegetarian diet. The chilly, damp climate made him very uncomfortable. He was forced by the weather to wear socks and shoes, but he hated them.

His chief escape was his work, and he poured himself into it with vigor, sometimes working 24 to 36 hours without rest. Ramanujan's brilliance, combined with Hardy's technical expertise, led to the publication of over 25 papers in European journals in five years. In 1918, Ramanujan was elected Fellow of the Royal Society of London and Fellow of Trinity College. He was the first Indian to receive either honor.

The Intriguing Taxicab

When Ramanujan was very ill, Hardy went to visit him in the hospital. It was an awkward moment. Not knowing what to say, Hardy made a comment about the ride over in the taxi.

"It was Taxi No. 1729. Rather a dull number, wouldn't you say?"

Ramanujan, weak as he was, propped himself up in the bed and smiled. "No, Hardy, not at all. 1729 is a fascinating number! It is the smallest number that can be expressed as the sum of two cubes in two different ways."

The Lost Notebook

In 1976, Ramanujan scholar George Andrews made a remarkable discovery. In a small box in the Trinity College library at Cambridge, he found an unbound manuscript of about 90 pages. Mathematical symbols were chaotically scribbled in all directions on both sides of the paper. It was the work of Ramanujan, apparently misplaced for nearly 50 years.

The notebook contains many mysterious formulas and complex equations. Today, mathematicians with sophisticated computers are proving Ramanujan's work in a breathtaking revelation of one of the greatest geniuses who ever lived.

COUNTING PARTITIONS

An amazing number theory discovery was made by the Indian mathematician, Ramanujan. He found a formula giving the exact number of partitions for any positive integer. No other mathematician thought this possible. A <u>partition</u> is a way of expressing an integer as the sum of other integers.

Consider a small number like 5. There are seven partitions possible, as shown.

$$5 = 1 + 1 + 1 + 1 + 1$$
$$5 = 2 + 1 + 1 + 1$$
$$5 = 2 + 2 + 1$$
$$5 = 3 + 1 + 1$$
$$5 = 3 + 2$$
$$5 = 4 + 1$$
$$5 = 5$$

1. Discover how many partitions there are for the number 3. List them.

2. List all the partitions for the number 4. How many are there?

3. There are 11 partitions for the number 6. See if you can find them.

4. List all the partitions for the number 7. Hint: you should find 15.

The number of partitions grows rapidly! For example, the number 100 has 190,569,292 partitions. The number 200 has approximately four trillion, or 3,972,999,029,388 partitions. You can see that finding a formula to derive the exact number of partitions was quite an amazing feat!

HISTORICAL CONNECTIONS VOL. II © 2005 AIMS Education Foundation

1729 GB

THE CURIOUS CAB

The untrained mathematical genius, Srinivasa Ramanujan, had an amazing ability to identify intricate number relationships.

Once, when Ramanujan was very ill, his colleague and friend G. H. Hardy took a taxi to the hospital to visit him. Hardy felt awkward and, trying to make conversation, said, "The taxi I rode in was No. 1729. Kind of a dull number, eh?"

Ramanujan propped himself up on his elbows and smiled. "No! 1729 is a wonderful number. It is the smallest number expressible as a sum of two cubes in two different ways."

Finding numbers that are the sum of one pair of cubes is easy. For example, $2^3 + 3^3 = 35$. But can you get 35 by summing two other cubes?

As Ramanujan said, the smallest number that can be stated as the sum of two pairs of cubes is 1729.

What are the two pairs of cubes for 1729?

To solve this problem, first make a list of cubes from 1 to 12.

Examine your list of cubes and find the two pairs Ramanujan knew about.

1729 = _____ + _____

1729 = _____ + _____

1^3 = _____ 1

2^3 = _____ 8

3^3 = _____

4^3 = _____

5^3 = _____

6^3 = _____

7^3 = _____

8^3 = _____

9^3 = _____

10^3 = _____

11^3 = _____

12^3 = _____

HISTORICAL CONNECTIONS VOL. II © 2005 AIMS Education Foundation

FINDING PI

Srinivasa Ramanujan discovered hundreds of amazing formulas. Some of his formulas are useful today in computing π(pi) to billions of decimal places.

Remember, π is the ratio of a circle's circumference to its diameter. In this activity you will have an opportunity to determine the value of π experimentally.

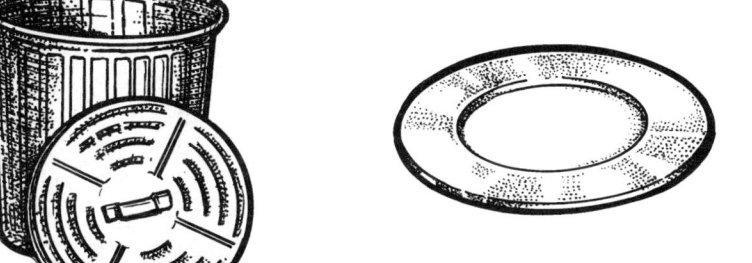

Collect circular objects such as garbage can lids, bicycle wheels, pie plates, or paper cups. Use a tape measure to measure the circumference and the diameter of each object. Record the measurements in the appropriate columns of this table. Complete the last column by dividing the circumference of each object by its diameter.

Object	Circumference (C)	Diameter (D)	$\frac{C}{D}$

Find the average of the values in the last column.

How does this compare with the known value of π?

NUMBERS: YOU GOTTA LOVE 'EM

Ramanujan, the great mathematician from India, is often called "the man who loved numbers." Many stories describe his love for and fascination with numbers.

Discover some surprising patterns in the number relationships in this activity.

Complete these tables, predicting numbers where there are blanks. Check your predictions by using a calculator.

1 x 9 + 1 = 10	1 x 8 + 1 = 9
12 x 9 + 2 = 110	12 x 8 + 2 = 98
123 x 9 + 3 = 1110	123 x 8 + 3 = 987
1234 x 9 + 4 = 11110	1234 x 8 + 4 = 9876

_____   _____
_____   _____
_____   _____
_____   _____

1 x 9109 = _____	1 x 1 = _____
2 x 9109 = _____	11 x 11 = _____
3 x 9109 = _____	111 x 111 = _____
4 x 9109 = _____	1111 x 1111 = _____
5 x 9109 = _____	11111 x 11111 = _____
6 x 9109 = _____	111111 x 111111 = _____
7 x 9109 = _____	1111111 x 1111111 = _____
8 x 9109 = _____	
9 x 9109 = _____	

FUN WITH NUMBERS

MAGIC DOMINOES

Here's a number trick you can play with friends.

Ask one person to select any domino from a domino set without letting you see it.

Then give your friend these instructions:
- Count the number of dots on the left side of the domino.
- Multiply this number by 5.
- Add 7.
- Multiply by 2.
- Add the number of dots on the right side of the domino.

Now, ask your friend to give you the total, and you can tell him or her which domino was selected.

Secret: Subtract 14 from the result.

The tens digit tells the number of dots on the left half of the domino and the ones digit tells the number on the right.

Can you explain why this works?

WHAT'S GOING ON HERE?

Take any number divisible by 3.   Example: 318

Add the next two larger numbers.   319
                                         320
                                         ___
                                         957

Add the digits of the sum.   9 + 5 + 7 = 21

Continue to add those digits.   2 + 1 = 3

You obtain the number 3, and you always will!

APPENDIX

THE NINE DIGIT PUZZLE

Cut out the cards below and arrange them in 3 columns to form a correct addition problem.

Find at least 30 different solutions. Record these on a sheet of paper.

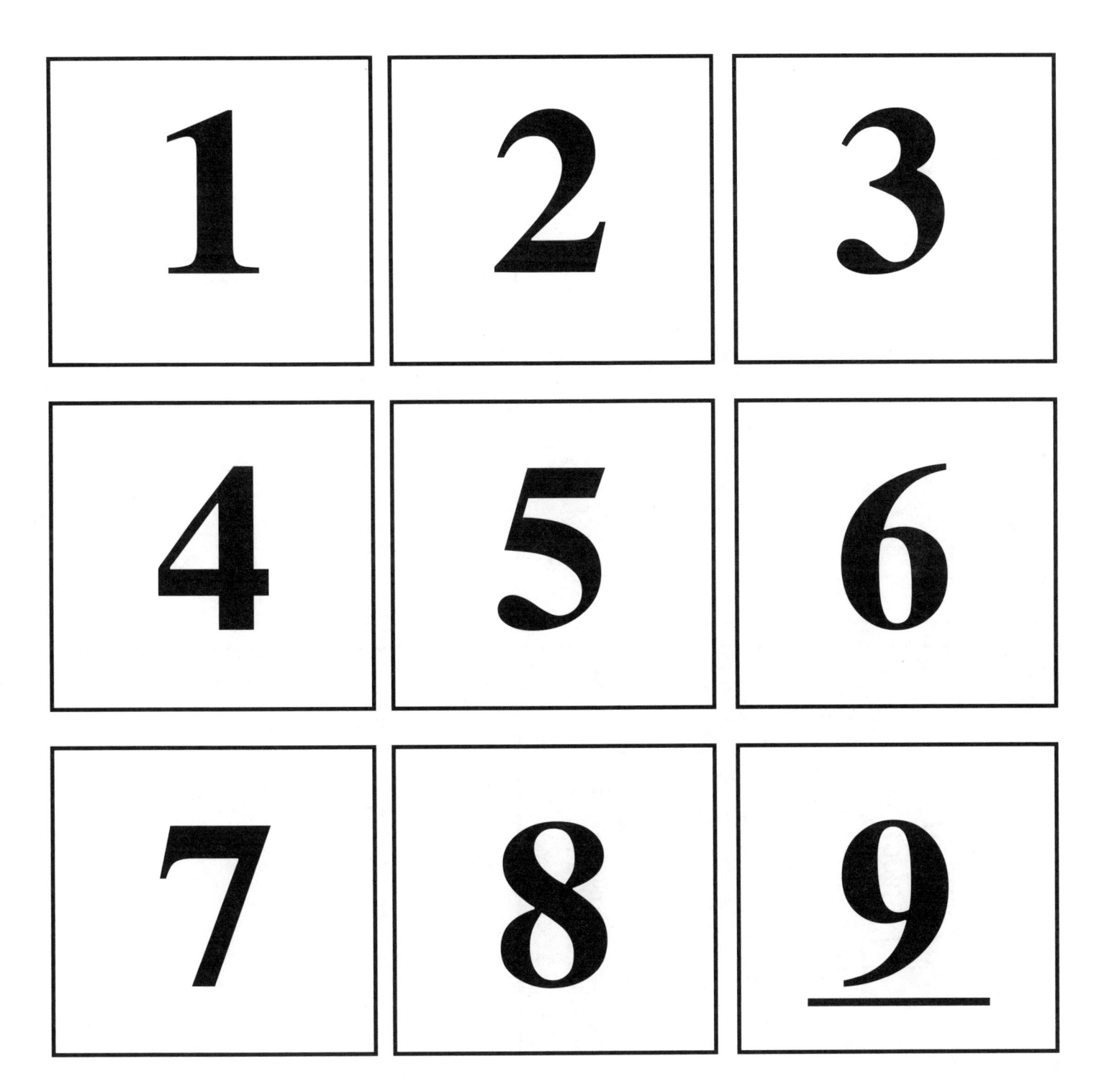

1	2	3
4	5	6
7	8	9

HISTORICAL CONNECTIONS VOL. II © 2005 AIMS Education Foundation

THE MAGIC CARD GAME

MAGIC CARD GAME INSTRUCTIONS

Ask someone to choose a number from 1-31 and hand you the cards on which that number appears. Quickly add the first number on each of those cards to determine the number your friend has chosen.

Example:
Suppose the player chooses the number 14. He/She must hand you cards B, C, and D. The first numbers on these cards are 2, 4, and 8. Their sum is 14, the selected number!

A
1	3	5	7
9	11	13	15
17	19	21	23
25	27	29	31

B
2	3	6	7
10	11	14	15
18	19	22	23
26	27	30	31

C
4	5	6	7
12	13	14	15
20	21	22	23
28	29	30	31

D
8	9	10	11
12	13	14	15
24	25	26	27
28	29	30	31

E
16	17	18	19
20	21	22	23
24	25	26	27
28	29	30	31

MAKING MAGIC CARDS

The integers 16, 8, 4, 2, and 1 can be added in various combinations to equal every number between 1 and 31. Using each integer only once, check which numbers you would add to get the numbers in the right hand columns.

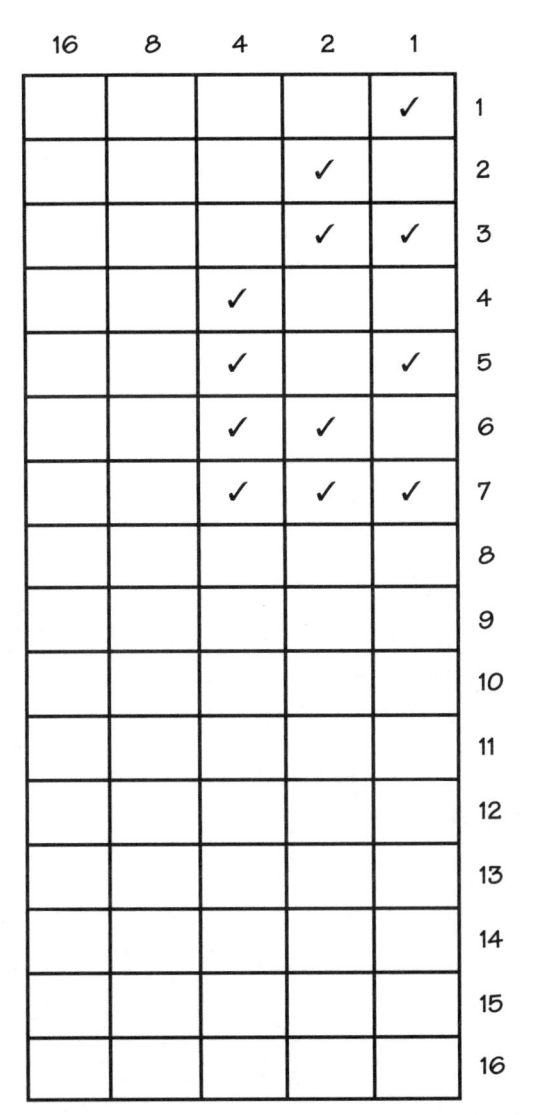

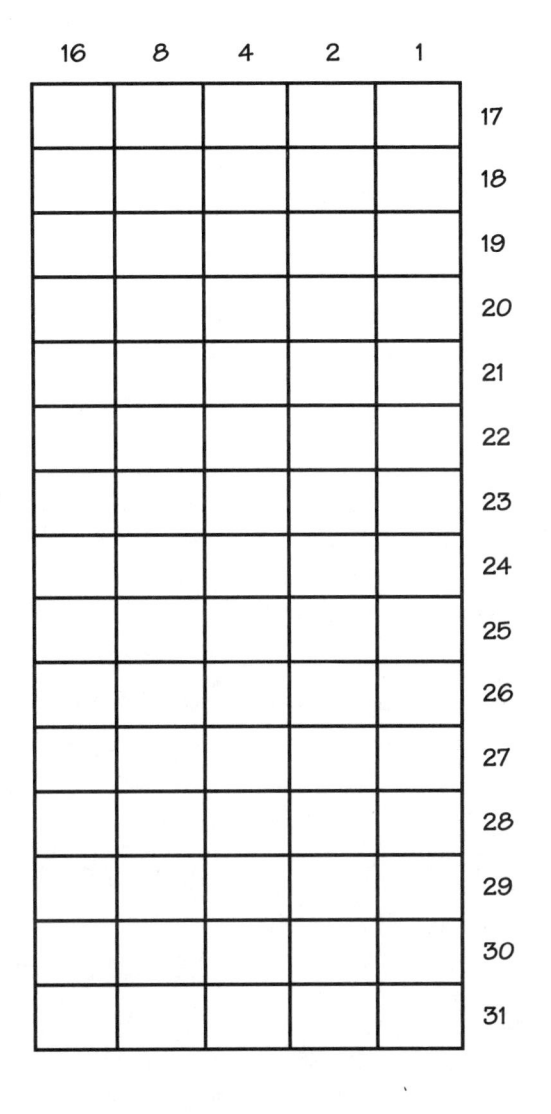

16	8	4	2	1	
				✓	1
			✓		2
			✓	✓	3
		✓			4
		✓		✓	5
		✓	✓		6
		✓	✓	✓	7
					8
					9
					10
					11
					12
					13
					14
					15
					16

16	8	4	2	1	
					17
					18
					19
					20
					21
					22
					23
					24
					25
					26
					27
					28
					29
					30
					31

Using the tables and check marks above, make your own magic cards. On Card A, write all the numbers with check marks in the 1 column. Place numbers with check marks in the 2 column on Card B, the 4 column on Card C, the 8 column on Card D, and the 16 column on Card E.

Each card should have 16 numbers when completed.

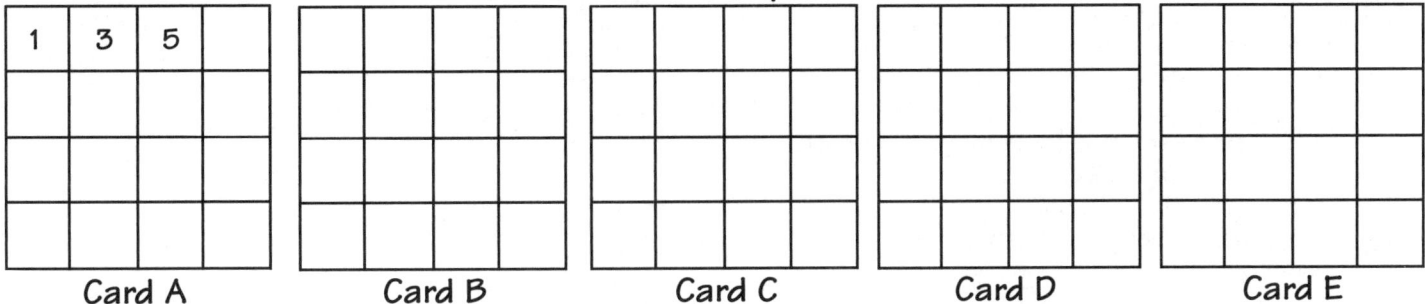

1	3	5	

Card A   Card B   Card C   Card D   Card E

HISTORICAL CONNECTIONS VOL. II   © 2005 AIMS Education Foundation

A CROSSWORD PUZZLE

ACROSS
1. Kovalevsky's home country.
6. Ramanujan created a remarkable _____ for finding pi.
7. Lovelace was a pioneer in computer _____.
8. Heron designed temple doors which opened using _____.
11. A sphere we can play with.
14. City where Hypatia lived.
15. Hypatia was murdered by an angry ____.
16. The portion of a curve between two points.
17. The first known woman mathematician.
21. Kovalevsky first learned calculus from her _____.
24. Galois was killed in a ____.
25. Ramanujan's home country.
28. Ramanujan returned to his home because he was ___.
29. The answer to an addition problem.
30. Kovalevsky could not study in Russia because she was a _____.
31. Babbage designed the first of these.
32. Babbage's home country.

DOWN
1. India's mathematical genius.
2. Where Banneker's grandfather was born.
3. A formula is sometimes called a ____.
4. Woman who worked with infinite series.
5. Banneker helped survey this city.
7. Thales measured these with shadows.
8. Heron found a method for finding _____ roots.
9. The computer language named for Lovelace.
10. The first computer programmer.
12. African-American mathematician.
13. Designed the first computer.
17. Triangle area formula is named for him.
18. What Banneker published.
19. The first named mathematician.
20. Young French mathematician not accepted by university.
22. Wrote *Elements*, best selling math text.
23. A positive integer with no factors other than 1 and itself is a _____ number.
24. Hypatia often engaged in a _____ with her students.
26. It carried salt for Thales.
27. Galois developed _____ theory.

MATHEMATICIANS
A CARD GAME

This fun, easy-to-play game will familiarize students with the names and contributions of 13 great mathematicians. More information and stories about them, plus related activities, can be found in Historical Connections, Volumes I and II.

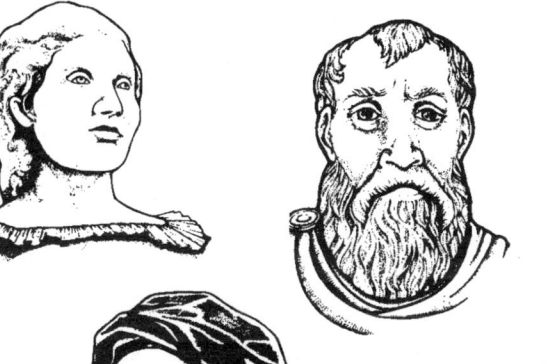

INSTRUCTIONS

MATHEMATICIANS contains 52 cards describing the contributions of 13 different mathematicians. The object of the game is to collect the four cards of a particular mathematician into a set. Two to six persons may play.

To begin the play, shuffle the cards and deal four to each person, placing the remainder face down within reach of all players. The player to the left of the dealer begins by asking any player for a particular trait of the mathematician he/she wants to collect. For example, a player may ask "Do you have Measured Pyramids Using Shadows'?" If the player has the card with this trait printed at the top, he/she must give it to the one who asked, who may then ask for another card from the same person or someone else. A player asks until he/she fails to get the card requested. The player then draws the top card from the deck. If the card drawn is the one just requested, he/she shows it to the group and keeps his/her turn. If not, the next person to the left takes a turn.

When a player gets a complete set, the set is placed face down in front of the player. When all the cards are in sets, the game ends. The player with the most sets wins.

HISTORICAL CONNECTIONS VOL. II   © 2005 AIMS Education Foundation

Card 1

Measured pyramids using shadows

Thales
c. 636 - c. 546 B. C.

One of "seven wise men" of Greece

First known mathematician

Strapped sponges on donkey to teach it a lesson

Card 2

One of "seven wise men" of Greece

Thales
c. 636 - c. 546 B. C.

First known mathematician

Strapped sponges on donkey to teach it a lesson

Measured pyramids using shadows

Card 3

First known mathematician

Thales
c. 636 - c. 546 B. C.

Strapped sponges on donkey to teach it a lesson

Measured pyramids using shadows

One of "seven wise men" of Greece

Card 4

Strapped sponges on donkey to teach it a lesson

Thales
c. 636 - c. 546 B. C.

Measured pyramids using shadows

One of "seven wise men" of Greece

First known mathematician

Card 1

Discovered $a^2 + b^2 = c^2$

Pythagoras
C. 560 - c. 480 B. C.

Founded secret group

Paid student to learn geometry

Said "Number rules the universe"

Card 2

Founded secret group

Pythagoras
C. 560 - c. 480 B. C.

Paid student to learn geometry

Said "Number rules the universe"

Discovered $a^2 + b^2 = c^2$

Card 3

Paid student to learn geometry

Pythagoras
C. 560 - c. 480 B. C.

Said "Number rules the universe"

Discovered $a^2 + b^2 = c^2$

Founded secret group

Card 4

Said "Number rules the universe"

Pythagoras
C. 560 - c. 480 B. C.

Discovered $a^2 + b^2 = c^2$

Founded secret group

Paid student to learn geometry

Card 1

Killed while doing geometry

Archimedes
287 - 212 B. C.

Invented weapons to fight off Romans

Solved crown problem for King Hieron

Invented screw to move water

Card 2

Invented weapons to fight off Romans

Archimedes
287 - 212 B. C.

Solved crown problem for King Hieron

Invented screw to move water

Killed while doing geometry

Card 3

Solved crown problem for King Hieron

Archimedes
287 - 212 B. C.

Invented screw to move water

Killed while doing geometry

Invented weapons to fight off Romans

Card 4

Invented screw to move water

Archimedes
287 - 212 B. C.

Killed while doing geometry

Invented weapons to fight off Romans

Solved crown problem for King Hieron

HISTORICAL CONNECTIONS VOL. II © 2005 AIMS Education Foundation

Card 1

First woman mathematician

Hypatia
c. 370 - c. 415

Respected for wisdom and beauty

Wrote about mathematics and astronomy

Murdered by an angry mob

Card 2

Respected for wisdom and beauty

Hypatia
c. 370 - c. 415

Wrote about mathematics and astronomy

Murdered by an angry mob

First woman mathematician

Card 3

Wrote about mathematics and astronomy

Hypatia
c. 370 - c. 415

Murdered by an angry mob

First woman mathematician

Respected for wisdom and beauty

Card 4

Murdered by an angry mob

Hypatia
c. 370 - c. 415

First woman mathematician

Respected for wisdom and beauty

Wrote about mathematics and astronomy

Card 1:

Discovered pendulum law

Galileo Galilei
1564-1642

Father of scientific method

Dropped objects from Leaning Tower of Pisa

16th century Italian mathematician

Card 2:

Father of scientific method

Galileo Galilei
1564-1642

Dropped objects from Leaning Tower of Pisa

16th century Italian mathematician

Discovered pendulum law

Card 3:

Dropped objects from Leaning Tower of Pisa

Galileo Galilei
1564-1642

16th century Italian mathematician

Discovered pendulum law

Father of scientific method

Card 4:

16th century Italian mathematician

Galileo Galilei
1564-1642

Discovered pendulum law

Father of scientific method

Dropped objects from Leaning Tower of Pisa

Card 1

Left famous "Last Theorem"

Pierre de Fermat
1601 - 1665

Did math in book margins

Father of number theory

Worked with Pascal on probability

Card 2

Did math in book margins

Pierre de Fermat
1601 - 1665

Father of number theory

Worked with Pascal on probability

Left famous "Last Theorem"

Card 3

Father of number theory

Pierre de Fermat
1601 - 1665

Worked with Pascal on probability

Left famous "Last Theorem"

Did math in book margins

Card 4

Worked with Pascal on probability

Pierre de Fermat
1601 - 1665

Left famous "Last Theorem"

Did math in book margins

Father of number theory

Card 1

A computer language is named for him

Blaise Pascal
1623 - 1662

Invented first mechanical calculator

Discovered properties of cycloid

Solved problems using Pascal's triangle

Card 2

Invented first mechanical calculator

Blaise Pascal
1623 - 1662

Discovered properties of cycloid

Solved problems using Pascal's triangle

A computer language is named for him

Card 3

Discovered properties of cycloid

Blaise Pascal
1623 - 1662

Solved problems using Pascal's triangle

A computer language is named for him

Invented first mechanical calculator

Card 4

Solved problems using Pascal's triangle

Blaise Pascal
1623 - 1662

A computer language is named for him

Invented first mechanical calculator

Discovered properties of cycloid

HISTORICAL CONNECTIONS VOL. II © 2005 AIMS Education Foundation

Card 1
A father of calculus

Isaac Newton
1642 - 1727

Discovered laws of motion

18th century British mathematician

Invented toys run by mice

Card 2
Discovered laws of motion

Isaac Newton
1642 - 1727

18th century British mathematician

Invented toys run by mice

A father of calculus

Card 3
18th century British mathematician

Isaac Newton
1642 - 1727

Invented toys run by mice

A father of calculus

Discovered laws of motion

Card 4
Invented toys run by mice

Isaac Newton
1642 - 1727

A father of calculus

Discovered laws of motion

18th century British mathematician

Card 1

Wrote more than 800 works

Leonhard Euler
1707 - 1783

Father of topology

First to use pi symbol (π)

Wasn't stopped by blindness

Card 2

Father of topology

Leonhard Euler
1707 - 1783

First to use pi symbol (π)

Wasn't stopped by blindness

Wrote more than 800 works

Card 3

First to use pi symbol (π)

Leonhard Euler
1707 - 1783

Wasn't stopped by blindness

Wrote more than 800 works

Father of topology

Card 4

Wasn't stopped by blindness

Leonhard Euler
1707 - 1783

Wrote more than 800 works

Father of topology

First to use pi symbol (π)

Card 1

First African-American mathematician

Benjamin Banneker
1731-1806

Computed math tables and almanacs

Built a striking clock out of wood

Helped survey Washington, D. C.

Card 2

Computed math tables and almanacs

Benjamin Banneker
1731-1806

Built a striking clock out of wood

Helped survey Washington, D. C.

First African-American mathematician

Card 3

Built a striking clock out of wood

Benjamin Banneker
1731-1806

Helped survey Washington, D. C.

First African-American mathematician

Computed math tables and almanacs

Card 4

Helped survey Washington, D. C.

Benjamin Banneker
1731-1806

First African-American mathematician

Computed math tables and almanacs

Built a striking clock out of wood

Card 1

Inspired by Archimedes' death

Sophie Germain
1776 - 1831

18th century French woman

Used man's name on work

Corresponded with Gauss on number theory

Card 2

18th century French woman

Sophie Germain
1776 - 1831

Used man's name on work

Corresponded with Gauss on number theory

Inspired by Archimedes' death

Card 3

Used man's name on work

Sophie Germain
1776 - 1831

Corresponded with Gauss on number theory

Inspired by Archimedes' death

18th century French woman

Card 4

Corresponded with Gauss on number theory

Sophie Germain
1776 - 1831

Inspired by Archimedes' death

18th century French woman

Used man's name on work

Card 1 (top-left):

Added 100 numbers in record time

Carl Gauss
1777 - 1855

The "Prince of Mathematics"

18th century German mathematician

One of the top three mathematicians

Card 2 (top-right):

The "Prince of Mathematics"

Carl Gauss
1777 - 1855

18th century German mathematician

One of the top three mathematicians

Added 100 numbers in record time

Card 3 (bottom-left):

18th century German mathematician

Carl Gauss
1777 - 1855

One of the top three mathematicians

Added 100 numbers in record time

The "Prince of Mathematics"

Card 4 (bottom-right):

One of the top three mathematicians

Carl Gauss
1777 - 1855

Added 100 numbers in record time

The "Prince of Mathematics"

18th century German mathematician

HISTORICAL CONNECTIONS VOL. II   © 2005 AIMS Education Foundation

Card 1
20th century mathematician from India

Srinivasa Ramanujan
1887 - 1920

Discovered many incredible formulas

Taught himself mathematics

Known for his love for numbers

Card 2
Discovered many incredible formulas

Srinivasa Ramanujan
1887 - 1920

Taught himself mathematics

Known for his love for numbers

20th century mathematician from India

Card 3
Taught himself mathematics

Srinivasa Ramanujan
1887 - 1920

Known for his love for numbers

20th century mathematician from India

Discovered many incredible formulas

Card 4
Known for his love for numbers

Srinivasa Ramanujan
1887 - 1920

20th century mathematician from India

Discovered many incredible formulas

Taught himself mathematics

HISTORICAL CONNECTIONS VOL. II © 2005 AIMS Education Foundation

SUGGESTIONS AND SOLUTIONS

Chapter One: Thales

Smart Shadows

The activity, "Smart Shadows," allows students to use Thales' technique to determine the height of flagpoles, trees, buildings, etc. The procedure is based on the proportion obtained from the two similar right triangles created by the shadows. For example, if the shadow cast by a 3 foot stick is 2 feet, and the shadow cast by a flagpole is 20 feet, the proportion is

$$\frac{3}{2} = \frac{\text{Height of pole}}{20}$$

$$\text{Height of pole} = \frac{(3)(20)}{2} = 30 \text{ ft.}$$

Stars Around The Moon

One of the important skills we can teach students is the ability to recognize and use patterns to solve problems. Thales was a good problem solver and solved many problems by observing patterns.

Solution: If a center dot (moon) exists in a square, count the surrounding dots (stars). Total these for each row.

Row totals: 6, 8, 6, 0, 0, 10, 20, 4, 4, 2, 10

Puzzling Pyramids

Students can have a lot of fun constructing these puzzles using practice golf balls and an electric glue gun. You will be surprised how intriguing these puzzles can be.

An activity entitled "Pyramid Puzzles" in Volume I of *Historical Connections in Mathematics* offers additional puzzles like these.

Dominoes on a Checkerboard

Thales is often called the father of deductive thinking, hence this deductive activity. Thirty-two dominoes can easily be placed on the 64 squares of a complete checkerboard. But when two opposite corners are removed from a checkerboard, the remaining squares cannot be covered with 31 dominoes. When a domino is placed on a checkerboard it must cover a white and a black square. When the opposite corners are removed, an unequal number of white and black squares remain, since the opposite corners are of the same color. It cannot be covered with 31 dominoes.

Who's Who on the Baseball Team?

This is a deductive activity where students must use the process of elimination to solve the problem.

Students should use the grid in the activity to check positions players cannot hold. For example, the clue, "Adam is a very close friend of the catcher," implies that Adam cannot be the catcher. Eventually students should see that the given clues allow for only one solution.

Eric:	Catcher
Jeff:	Pitcher
Troy:	1st Base
Paul:	2nd Base
Adam:	3rd Base
Matt:	Shortstop
Luis:	Left field
Jose:	Center field
Lee:	Right field

Chapter Two: Euclid

Euclid's Algorithm

The Euclidean algorithm is a powerful process for finding the greatest common divisor (GCD) of two positive integers. This algorithm has important applications in various areas of mathematics, including Diophantine equations and continued fractions.

This activity will familiarize students with the process used in Euclid's algorithm. The algorithm may also be used to reduce fractions. Most students find the process intriguing.

Some teachers may wish to extend the activity to finding the GCD for more than two numbers. This can easily be done by first finding the GCD for two numbers, then using the result with the third, etc.

Solutions:
1a. 2
1b. 39
1c. 12
1d. 12
1e. 60
1f. 67

2a. $\frac{8}{23}$ 2b. $\frac{3}{7}$

2c. $\frac{3}{5}$ 2d. $\frac{3}{7}$

Greatest Common Divisor Meets The Least Common Multiple

This activity allows students to discover an interesting and widely unknown relationship between two numbers and their GCD and LCM. The relationship can be expressed in a number of ways, one of which is A X B = GCD X LCM. Allow students to express other variations of this such as $\frac{B}{GCD} = \frac{LCM}{A}$ etc. With this discovery students are challenged to complete the lower part of the table.

A	B	GCD	LCM
8	12	4	24
6	40	2	120
9	30	3	90
24	36	12	72
15	25	5	75
18	4	2	36
Use what you discovered above to complete the table.			
33	22	11	66
120	50	10	600
38	56	2	1064
620	280	20	8680

The following BASIC program will find the greatest common divisor (GCD) and the least common multiple (LCM) for the two numbers entered. It will accept any two numbers between 1 and 1 billion. It uses Euclid's Algorithm and is extremely fast.

BASIC PROGRAM FOR GCD AND LCM:

```
10   HOME
20   PRINT "THIS PROGRAM WILL FIND THE
     GCD AND LCM FOR THE TWO NUMBERS
     YOU ENTER."
30   PRINT
40   INPUT "ENTER FIRST NUMBER: ";A
50   INPUT "ENTER SECOND NUMBER: ";B
60   P= A*B
70   IF A<=0 THEN 220
80   IF B<=0 THEN 220
90   IF A>=B THEN 130
100  C=A
110  A=B
120  B=C
130  D=A-B*INT(A/B)
140  A=B
150  B=D
160  IF D<>0 THEN 130
170  PRINT
180  PRINT "THE GCD IS "A
190  PRINT "THE LCM IS "P/A
200  PRINT
210  GOTO 30
220  PRINT "PLEASE! POSITIVE INTEGERS
     ONLY!"
230  GOTO 30
240  END
```

Perfect Numbers

Perfect numbers have intrigued people from earliest days. In this activity, students will discover a procedure for finding perfect numbers.

1st Column	2nd Column	Perfect Numbers
2	3	6
4	7	28
8	15	
16	31	496
32	63	
64	127	8128
128	255	
256	511	
512	1023	
1024	2047	
2048	4095	
4096	8191	33,550,336

Daffy Definitions from Geometry

Euclid is often called the father of geometry. As students begin studying geometry, they are exposed to many geometric definitions, some familiar and some new.

This is a light-hearted activity allowing students to test their creativity and to build familiarity with some geometric terms.

1. Right angle
2. Chord
3. Complement
4. Arc
5. Center
6. Degree
7. Acute angle
8. Polygon
9. Inverse
10. Pi
11. Protractor
12. Tangent
13. Rectangle
14. Ruler
15. Geometry ("Gee, I'm a Tree")
16. Hypotenuse (High pot in use)
17. Cone
18. Parallelogram
19. Trapezoid
20. Ratio
21. Circumference ("Sir, come for ants.")
22. Line
23. Plane
24. Coincide (Go inside)
25. Secant

White-Faced Cubes

A model such as a Rubik's cube may help students visualize this problem.

In this activity, students often begin thinking inductively (looking for the vertical pattern in each column) and then suddenly switch to deductive thinking by noticing where the various smaller cubes are located on the big cube–on an edge, in the middle of a face, on a corner, etc.

Length of edge	Number of Cubes formed	Number with 3 white faces	Number with 2 white faces	Number with 1 white face	Number with no white faces
2	8	8	0	0	0
3	27	8	12	6	1
4	64	8	24	24	8
5	125	8	36	54	27
6	216	8	48	96	64
20	8000	8	216	1944	5832
n	n^3	8	$12(n-2)$	$6(n-2)^2$	$(n-2)^3$

Chapter Three: Heron

Heron's Square Root Method

1. $\sqrt{37}$
   First approximation: 6
   Second approximation: $\frac{1}{2}(6 + \frac{37}{6}) = \frac{73}{12} = 6.08$

2. $\sqrt{27}$
   First approximation: 5
   Second approximation: $\frac{1}{2}(5 + \frac{27}{5}) = \frac{52}{10} = 5.20$

3. $\sqrt{14}$
   First approximation: 4
   Second approximation: $\frac{1}{2}(4 + \frac{14}{4}) = \frac{30}{8} = 3.75$

4. $\sqrt{87}$
   First approximation: 9
   Second approximation: $\frac{1}{2}(9 + \frac{87}{9}) = \frac{168}{18} = 9.33$

Heron's Formula

Solutions:
1. The area of the two triangles is the same. Each has an area of 300 square units.

2. (a) $A = \sqrt{6(6-3)(6-4)(6-5)} = 6$

   (b) $A = \frac{(3)(4)}{2} = 6$

3. $A = \sqrt{34(17)(9)(8)} = 204$ square units

The Chocolate Cake Challenge

This problem has a very surprising solution. Most people do not believe that simply dividing the perimeter into equal parts and then cutting from the center of the cake is the correct solution.

To show that cake and frosting are equal for each piece, find the area of several regions. Some corner pieces need to be divided into two triangles for the computation.

Space for Ace

Students must be reminded to think carefully about this situation. Ace can roam over $\frac{3}{4}$ of a circle of radius 60 feet and $\frac{1}{4}$ of a circle of radius 20 feet.

Hence, the total area is $(\frac{3}{4})(\pi)(60^2) + (\frac{1}{4})(\pi)(20^2) = 8792$ square feet using 3.14 for π.

Answers will vary slightly depending on value used for π.

Diagonal Challenge

Remind students that a diagonal of a polygon is a line segment joining two non-adjacent vertices.

It is easiest to count the diagonals as they are drawn. To be sure all possible diagonals are drawn, have students examine how many are drawn from each vertex of the polygon. This should always be three fewer than the number of sides.

Students should notice that the differences for the numbers in the second column are 2, 3, 4, 5, etc. This discovery allows students to continue the table indefinitely.

There are several ways to obtain the solution for n sides. Increasing each number in the second column by one results in the triangular numbers. Thus, a polygon with n sides has $\frac{(n-2)(n-1)}{2}$ diagonals. This expression simplifies to $\frac{n(n-3)}{2}$.

Number of Sides	Number of Diagonals
3	0
4	2
5	5
6	9
7	14
8	20
9	27
10	35

Chapter Four: Hypatia

The Amazing Ellipse

Hypatia dedicated much of her writing and teaching to an understanding of conic sections, of which the ellipse and parabola are two. In this activity, students will learn how to draw an ellipse and discover some of the properties of the ellipse.

Displaying ellipses with different focal points will show that as the focal points are placed closer to the center, the ellipse approaches a circle.

The area of an ellipse is given by the formula πab where a and b are the lengths of the semi-major and semi-minor axes (half the longer and half the shorter axes). When the ellipse becomes a circle, a and b are equal and πab becomes πa^2, the well-known formula for the area of a circle.

Teachers may share another application: Imagine a pool or billiard table in the shape of an ellipse. If a ball were placed at one focus and hit in any direction, it would bounce off the wall and pass through the other focus. In the case of the whispering gallery, all sound waves coming from one focus bounce off the wall and pass through the other focus.

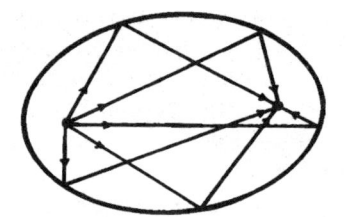

Changing a Quarter

Hypatia was interested in and wrote about Diophantine problems, of which "Changing a Quarter" is a modern day version.

Diophantus gave algebraic methods for finding all solutions to this kind of problem. In this activity, students are asked to find the solutions experimentally, watching for any patterns that might emerge.

Number of Dimes	Number of Nickels	Number of Pennies
2	1	0
2	0	5
1	3	0
1	2	5
1	1	10
1	0	15
0	5	0
0	4	5
0	3	10
0	2	15
0	1	20
0	0	25

Super challenge solution:
There are 292 ways to make change for one dollar.

Paper Folding a Parabola

The parabola is another conic section. It is defined as the set of all points in a plane such that the distance of each point from a fixed point is the same as its distance from a fixed line. The fixed point is called the *focus* of the parabola.

There are many interesting applications of the parabola such as those mentioned on the activity sheet. The paper folding activity is a simple, interesting method for students to create their own parabolas. The students should make about 20 folds, folding the paper in a different direction each time.

This activity may be effectively demonstrated on an overhead projector by folding waxed paper. It might be interesting to display some of the students' parabolas.

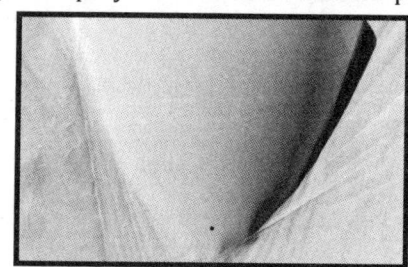

A Stamp Stumper
This is another example (see "Changing a Quarter") of a Diophantine problem. As students find various solutions, they should watch for emerging patterns.

Number of 3-cent stamps	Number of 8-cent stamps
0	25
8	22
16	19
24	16
32	13
40	10
48	7
56	4
64	1

Chapter Five: Banneker

Two Puzzles from Benjamin Banneker's Collection
Banneker enjoyed collecting and solving puzzles. This activity features two from his collection, updated for today's students.

The Ladder Puzzle: 44 feet
If students need help, suggest they use the Pythagorean Theorem.

$\sqrt{50^2 - 48^2}$ = 14 ft

$\sqrt{50^2 - 40^2}$ = 30 ft

14 ft + 30 ft = 44 ft

The Cows, Goats, and Chickens Puzzle: 19 cows, 1 goat, 80 chickens

A Clockmaker's Challenge
This activity illustrates in a small way the challenge Banneker faced in constructing his famous striking clock, carving the gears out of wood.
Solutions:
1a. Counter-clockwise
1b. Clockwise
1c. Two
1d. Four
2a. Counter-clockwise
2b. Four
2c. Sixteen

A Collection of Puzzlers
Banneker loved puzzles of all kinds. He often copied puzzles about farm animals and household objects into his journal so that he could enjoy them again and again.

While these should not be considered "trick" questions, students should be encouraged to think carefully before concluding their answers are correct.
Solutions:
1. 59 minutes
2. 3 inches
3. They are triplets.
4. 28 hours
5. 20 miles
6. 12¢
7. His son
8. Three
9. After 14 days
10. 20 feet

Make-A-Square Puzzle
This puzzle helps students to visualize the Pythagorean Theorem. The activity works best when photocopied onto light card stock. Teachers may wish to make a wooden model as well. Another Pythagorean puzzle may be found in *Historical Connections in Mathematics* Volume I, Chapter 1.

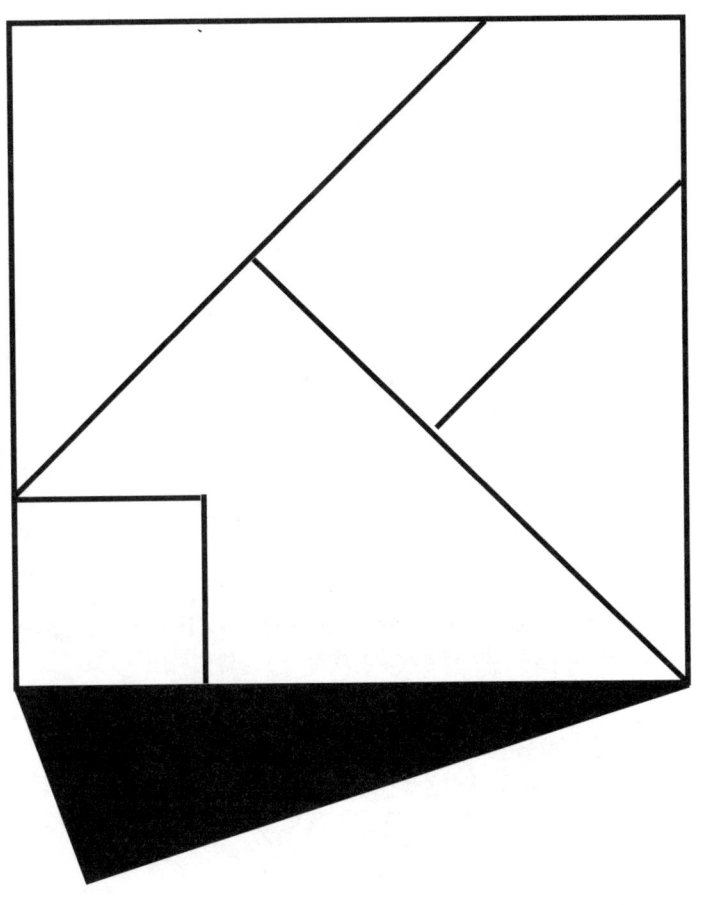

A Jumble Puzzle
This puzzle uses a common format to reinforce some facts about Banneker's life.

Jumbles:
    WATCH
    POCKET
    BOOKS
    PUZZLE
    MARYLAND

Answer: A WOODEN CLOCK

Chapter Six: Babbage

All of the activities in this chapter relate to the method of differences Babbage used in designing the first computer called the *difference engine*.

When the first few rows of a table were entered into Babbage's computer, it would figure out the pattern using differences between the numbers in the second column and then run the table indefinitely.

Rings of Circles

This activity shows students the problem-solving power in Babbage's use of differences, commonly known as the method of finite differences. Looking at differences often provides the "clue" required to continue a pattern or formulate a generalization.

Students should be allowed to experiment with techniques to complete the table. At first, they may count the circles in each array. Eventually, they should try to discover the pattern using differences.

Note that the size of the circles has no impact on the number required, but they must be uniform in each design.

0	1
1	7
2	19
3	37
4	61
5	91
6	127
7	169
8	217

Challenge students to find the formula for n rings in the "Rings of Circles" activity. The solution is
$3n^2 + 3n + 1$

Hint: Subtract 1 from each number in the second column. Note that the results are all divisible by 6.

A colorful bulletin board display can be created by glueing circular colored chips to card stock in "Rings of Circles" designs.

Pizza Party

In this activity, encourage students to be careful how the successive cuts are made to assure the maximum number of pieces. Each new cut must cross all the previous cuts, but it must not pass through any existing intersection points. Upon examining the differences in the numbers in the second column, we obtain 2, 3, 4, 5, 6, etc., and the pattern becomes obvious.

Number of Cuts	Maximum Number of Pieces
1	2
2	4
3	7
4	11
5	16
6	22
7	29
8	37

The formula for n cuts may be obtained by reducing each number in the second column by one. The result is the triangular numbers. The solution for n cuts is $\frac{n(n+1)}{2} + 1$.

Tournament Time

This problem gives students practice setting up "tournament" diagrams and determining the relationship between the number of teams in a single elimination tournament and the total number of games played. The solution is shown below.

Number of Teams	Number of Games
2	1
3	2
4	3
5	4
6	5
7	6
8	7
9	8
10	9
n	n - 1

Region Revenge

This activity illustrates the power of the difference method utilized by Babbage in his design for the first computer. The numbers in the second column will surprise students in that the pattern goes 1, 2, 4, 8, 16, 31, and not 32. Remind students to count carefully and then look at differences, difference of differences, etc., to uncover the vertical pattern so they can complete the table.

Number of Points	Maximum Number of Regions
1	1
2	2
3	4
4	8
5	16
6	31
7	57
8	99
9	163
10	256

Checkerboard Challenge

Students may be encouraged to organize their counting by counting all 1 by 1 squares, then 2 by 2 squares, etc. The solution may be found by looking at the differences between the numbers in the second column.

1 by 1	1
2 by 2	5
3 by 3	14
4 by 4	30
5 by 5	55
6 by 6	91
7 by 7	140
8 by 8	204

Differences: 4, 9, 16, 25, 36, 49, 64

Chapter Seven: Galois

Rotating Tires: A Mathematical Event!

This activity introduces students to *mathematical groups*. Briefly, a group is a set of elements with a rule showing how these elements can be combined. This rule is often given in the form of a table. A group must satisfy several additional properties that are not discussed here.

Solution:

*	0	1	2	3
0	0	1	2	3
1	1	0	3	2
2	2	3	1	0
3	3	2	0	1

Clock Arithmetic

This activity introduces students to another type of addition than they commonly study. Clock arithmetic is also an example of a mathematical system called a group. Galois laid much of the ground work for what is called group theory.

+	1	2	3	4	5	6	7	8	9	10	11	12
1	2	3	4	5	6	7	8	9	9	11	12	1
2	3	4	5	6	7	8	9	10	11	12	1	2
3	4	5	6	7	8	9	10	11	12	1	2	3
4	5	6	7	8	8	10	11	12	1	2	3	4
5	6	7	8	9	10	11	12	1	2	3	4	5
6	7	8	9	10	11	12	1	2	3	4	5	6
7	8	9	10	11	12	1	2	3	4	5	6	7
8	9	10	11	12	1	2	3	4	5	6	7	8
9	10	11	12	1	2	3	4	5	6	7	8	9
10	11	12	1	2	3	4	5	6	7	8	9	10
11	12	1	2	3	4	5	6	7	8	9	10	11
12	1	2	3	4	5	6	7	8	9	10	11	12

1. 1 p.m.

2. 4 p.m.

Magic Flexagons

Students will be fascinated with this activity and will enjoy making their own tri-hexa-flexagon. Some teachers may wish to enlarge the pattern on a photocopy machine.

Flexagons began with a British graduate student named Stone who was studying at Princeton. It seems that American notebook paper was too large for his British notebook, so he trimmed off strips. He began to play with these strips, folding them into curious shapes, and discovered the amazing flexagon.

The "how to flex" section should help them "break in" a new flexagon. Have them flex them several times in succession. Emphasize the importance of firmly creasing the folds. If a student's flexagon will not open, try the alternate vertices.

When completing the table, remind students to always start with the flexagon in its initial position of GB (gray on top, black on bottom).

Solution:
1. 5317 LIES
2. 8075 SLOB
3. 7735 SELL
4. 35108 BOISE
5. 5733 EELS
6. 7738 BELL
7. 3705 SOLE
8. 7105 SOIL
9. 5508 BOSS
10. 5338 BEES
11. 35007 LOOSE
12. 57718 BILLS

Circle Circus

This activity gives students the opportunity to discover the relationship between overlapping circles and the maximum number of interior regions created by these circles.

Number of Circles	Maximum Number of Regions
1	1
2	3
3	7
4	13
5	21
6	31
7	43
8	57

After students have entered the data in the table for the circles shown, they should discover that the differences of the numbers in the second column are even numbers, 2, 4, 6, 8, etc. This discovery will allow them to complete the table.

To help students discover the solution for n circles, suggest they reduce each number in the second column by one. It is now easier to see the relationship between these new numbers and the numbers in the first column. The number of regions for n circles is $n(n-1) + 1$, which can also be expressed as $n^2 - n + 1$.

Twenty-One Connect

This problem demonstrates the power of solving a problem by examining a number of simple cases, discovering the emerging pattern, and generalizing.

Students should count the number of segments for circles with 5 points and 6 points. To be sure all possible segments are drawn, have students examine how many originate from each point. This should always be one fewer than the number of points on the circle.

"The Erased Chance": A Skit to Read

This skit portrays only one of the many unfortunate events in Galois' life. After class reading, students might wish to discuss how much of Galois' trouble was his own fault. While it is true that he was often misunderstood and under-rated, and suffered plain bad luck, it is also apparent that he bore some responsibility. How might Galois have responded more productively in this particular situation? Other students may wish to talk about why (or if) showing one's work is important.

Chapter Eight: Lovelace

Ada Lovelace worked with Charles Babbage on his difference and analytical engines. She is often considered the first computer programmer because she determined how to program the analytical engine. The activities in this chapter relate to differences, calculators, and computers.

Calculator Fun

While Lovelace is known as the first programmer, Blaise Pascal (1623-62) invented the first mechanical calculator.

Note that the differences for the numbers in the second column are 1, 2, 3, 4, 5, etc. This discovery allows students to continue the table indefinitely. If students are familiar with triangular numbers, they will recognize the numbers in the second column as triangular numbers.

Number of Points	Number of Segments
2	1
3	3
4	6
5	10
6	15
7	21
21	210
n	$\frac{n(n-1)}{2}$

Leap Frog

This activity can be done using colored chips or pennies and dimes on the grid at the bottom of the activity sheet. A small board with 9 holes to accomodate a total of 8 golf tees (4 of one color and 4 of another) also works very nicely.

Allow students to learn how to make the moves according to the rules. Be patient. Some will catch on faster than others.

After mastering the moves, students should enter the data in the table and watch for the emerging pattern. Data should be collected for up to four pegs on each half. Extend the grid at the bottom of the activity sheet if necessary.

Number of Pegs on One Half	Minimum Number of Moves
1	3
2	8
3	15
4	24
5	35
20	440
n	$(n+1)^2 - 1$ or $n^2 + 2n$

If students need a hint to find the solution for n pegs, suggest they increase each number in the second column by one. This yields the perfect squares.

Some students may discover another method. Multiply the number in the first column by a number two larger.

Both generalizations, of course, are algebraically the same.

A good extension of the "Leap Frog" activity is to determine how many of the moves are jumps, that is, moves where one jumps over an opposite color.

The table for this is shown.

Number of Pegs on One Half	Number of Jumps
1	1
2	4
3	9
4	16
5	25
20	400
n	n^2

Note that in the formula for the total number of moves, $n^2 + 2n$, n^2 represents the number of jumps; therefore, $2n$ must represent the number of non-jumps.

Some teachers have made 6 foot models of this activity with approximately 40 golf tees on each half. They present the problem showing the large model and asking how many moves would be required. The process leads to looking at simpler situations first, discovering the emerging pattern, and then predicting for 40. It's quite dramatic!

Round Robin Play

Many students are familiar with round robin play since many schools use this system when playing other schools in a league.

Solution:

Number of Teams	Number of Games
2	1
3	3
4	6
5	10
6	15
7	21
8	28
n	$\frac{n(n-1)}{2}$

HISTORICAL CONNECTIONS VOL. II

Chapter Nine: Kovalevsky

The Snowflake Curve

The activity entitled "The Snowflake Curve" illustrates infinite sequences, which were of great interest to Sonya Kovalevsky and other nineteenth-century mathematicians. As the snowflake "grows," the perimeter becomes infinitely large (in mathematical terminology, the perimeter sequence is said to be *divergent*) whereas the area approaches a finite quantity (the area sequence is *convergent*).

Solutions:

Stage Number	Number of Sides	Length of each side (in inches)	Perimeter of Snowflake (in inches)
1	3	1	$3 \cdot 1 = 4$
2	12	$\frac{1}{3}$	$12 \cdot \frac{1}{3} = 4$
3	48	$\frac{1}{9}$	$48 \cdot \frac{1}{9} = 5\frac{1}{3}$
4	192	$\frac{1}{27}$	$192 \cdot \frac{1}{27} = 7\frac{1}{9}$
5	768	$\frac{1}{81}$	$768 \cdot \frac{1}{81} = 9\frac{13}{27}$
n	$3 \cdot 4^{n-1}$	$\frac{1}{3^{n-1}}$	$\frac{4^{n-1}}{3^{n-2}}$

Some students may need help creating the formulas for the nth stage. Teachers may give appropriate hints such as "Divide each number in the 'Number of Sides' column by 3. The resulting numbers are powers of 4."

$$P(60) = \frac{4^{59}}{3^{58}} \text{ inches} = 1114 \text{ miles}$$

$$P(100) = \frac{4^{99}}{3^{98}} \text{ inches} = 110{,}723{,}807 \text{ miles}.$$

Some may wish to extend this activity by deriving the *area* of the snowflake. For the nth stage, this formula is

$$A = \frac{2\sqrt{3}}{5} - \frac{3\sqrt{3}}{20}\left(\frac{4}{9}\right)^{n-1}$$

Find the area (in square inches) for stage 1, stage 3, and stage 100.

Answers:
A(1) = .433 square inch
A(3) = .642 square inch
A(100) = .693 square inch.

The area approaches $\frac{2\sqrt{3}}{5}$ square inch as n increases indefinitely.

For a dramatic classroom presentation, show the snowflake curve being drawn on a computer screen.

Enter the following Apple Logo programs for the snowflake curve.

```
TO SIDE :SIZE :STAGE
IF :STAGE = 1 [FD :SIZE STOP]
SIDE :SIZE/3 :STAGE -1
LT 60
SIDE :SIZE/3 :STAGE -1
RT 120
SIDE :SIZE/3 :STAGE -1
LT 60
SIDE :SIZE/3 :STAGE -1
END

TO SN :STAGE
MAKE "SIZE 180
PU SETPOS [-50 -75]
PD
REPEAT 3[SIDE :SIZE :STAGE RT 120]
END
```

To run the program, enter SN followed by a number that indicates the stage to be drawn. For example, when SN 1 is entered, the turtle will draw stage 1 of the snowflake curve. When SN 2 is entered, stage 2 is drawn, etc. Draw several stages, clearing the screen (CS) between drawings.

Predictable Patterns

Sonya Kovalevsky contributed to the study of infinite series and sequences. Seeing patterns and forming generalizations is an important part of this work. In this activity students have the experience of making discoveries and predicting some number sequences.

Solutions:
1. 11, 13, 15, 17
2. 24, 28, 32
3. 16, 22, 29, 37
4. 21, 28, 36
5. 36, 49, 64, 81
6. 32, 64, 128, 256
7. 15, 18, 21, 24
8. 42, 41, 123, 122 (subtract 1, times 3)
9. 5, 4, 6, 5
10. 57, 93 (increase by squares)
11. 34, 55, 89, 144 (Fibonacci sequence)

12. 45, 59, 75

13. 37, 50, 65

14. 8, 1, 1, 0 (square numbers with commas inserted)

15. 36, 21, 49, 28 (square and triangular numbers alternate)

The Chessboard Covered With Wheat

The result of this problem surprises almost everyone. The power of two sequence is very explosive; it increases much faster than most people imagine.

Number of Squares	Total Number of Grains
1	1
2	3
3	7
4	15
5	31
6	63
64	$2^n - 1$

Paper Punching Patterns

This hands-on activity gives rise to an interesting and surprising sequence. It may work especially well with students in small groups.

Number of times folded in half	Number of holes
0	1
1	3
2	7
3	15
4	31
5	63
6	127
7	255
8	511
20	2,097,151
n	$2^{n+1} - 1$

Chapter Ten: Ramanujan

Counting Partitions

The activity, "Counting Partitions," introduces students to the concept of partitions, important in mathematics, communications and modern physics.

Solutions:
1. 3 = 1+1+1
   3 = 2+1
   3 = 3

2. 4 = 1+1+1+1
   4 = 2+1+1
   4 = 2+2
   4 = 3+1
   4 = 4

3. 6 = 1+1+1+1+1+1
   6 = 2+1+1+1+1
   6 = 2+2+1+1
   6 = 2+2+2
   6 = 3+1+1+1
   6 = 3+2+1
   6 = 3+3
   6 = 4+1+1
   6 = 4+2
   6 = 5+1
   6 = 6

4. 7 = 1+1+1+1+1+1+1
   7 = 2+1+1+1+1+1
   7 = 2+2+1+1+1
   7 = 2+2+2+1
   7 = 3+1+1+1+1
   7 = 3+2+1+1
   7 = 3+2+2
   7 = 3+3+1
   7 = 4+1+1+1
   7 = 4+2+1
   7 = 4+3
   7 = 5+1+1
   7 = 5+2
   7 = 6+1
   7 = 7

Encourage students to list partitions in a systematic way and compare their methods with others'.

Some teachers may wish to use the following BASIC program to find the number of partitions for each of the first 80 integers.

```
10  HOME
20  PRINT "THIS PROGRAM FINDS THE NUMBER OF
    PARTITIONS FOR THE FIRST 80 INTEGERS."
30  PRINT "BE PATIENT!"
40  PRINT
50  DIM P(80,80), Q(80)
```

HISTORICAL CONNECTIONS VOL. II   113   © 2005 AIMS Education Foundation

```
60   LET U = 80
70   FOR I = 1 TO U
80   LET P(I,I) = 1
90   LET Q(I) = 1
100  NEXT I
110  PRINT "N", "NO. OF PARTITIONS OF N"
120  POKE 34,5
130  PRINT "1", "1"
140  FOR I = 2 TO U
150  FOR J = 1 TO I – 1
160  FOR K = 1 TO J
170  LET P(I,J) = P(I,J) + P(I–J, K)
180  NEXT K
190  LET Q(I) = Q(I) + P(I,J)
200  NEXT J
210  PRINT I, Q(I)
220  NEXT I
230  END
```

The Curious Cab

This activity is based on the story told most often about Ramanujan. After completing the table of cubes, students should find the two pairs of cubes Ramanujan knew about are

$1729 = 1^3 + 12^3$ or $1 + 1728$ and

$1729 = 9^3 + 10^3$ or $729 + 1000$

1^3	=	1
2^3	=	8
3^3	=	27
4^3	=	64
5^3	=	125
6^3	=	216
7^3	=	343
8^3	=	512
9^3	=	729
10^3	=	1000
11^3	=	1331
12^3	=	1728

Finding Pi

This activity gives students an opportunity to approximate the value of π. Measuring the circumference and diameter of circular objects and finding the ratio will reinforce the meaning of π as few other experiments can.

Numbers—You Gotta Love 'Em

Ramanujan, the great Indian mathematician, is often called "the man who loved numbers." In this activity students will discover some interesting number relationships.

1	x 9	+ 1	=	10
12	x 9	+ 2	=	110
123	x 9	+ 3	=	1110
1234	x 9	+ 4	=	11110
12345	x 9	+ 5	=	111110
123456	x 9	+ 6	=	1111110
1234567	x 9	+ 7	=	11111110
12345678	x 9	+ 8	=	111111110

1	x 8	+ 1	=	9
12	x 8	+ 2	=	98
123	x 8	+ 3	=	987
1234	x 8	+ 4	=	9876
12345	x 8	+ 5	=	98765
123456	x 8	+ 6	=	987654
1234567	x 8	+ 7	=	9876543
12345678	x 8	+ 8	=	98765432

1	x 9109	=	9109
2	x 9109	=	18218
3	x 9109	=	27327
4	x 9109	=	36436
5	x 9109	=	45545
6	x 9109	=	54654
7	x 9109	=	63763
8	x 9109	=	72872
9	x 9109	=	81981

1 x 1	=	1
11 x 11	=	121
111 x 111	=	12321
1111 x 1111	=	1234321
11111 x 11111	=	123454321
111111 x 111111	=	12345654321
1111111 x 1111111	=	1234567654321

Fun With Numbers

Students are always interested in magic. "Magic Dominoes" and "What's Going On Here?" capitalize on that interest. They may enjoy the activity (and practice computation) without understanding how it works, but other students will accept the challenge of understanding the "trick."

Here's how "Magic Dominoes" works:

Let x represent the number of dots on the left side of the domino and y the number on the right.

Following the instructions in the activity, we get

5x

5x + 7

2(5x + 7) = 10x + 14

10x + 14 + y

Subtracting 14 leaves 10x + y, which is the value of a two-digit number where x is the tens digit and y is the units digit.

Appendix

The Nine-Digit Puzzle

The object of this puzzle is to arrange the nine pieces into a three digit addition problem with the correct sum.

For example,

271	418	467	158
683	275	352	634
954	693	819	792

Allow students to find arrangements that work, and record them on the board. There are over 300 possible solutions!

Challenge students to discover the key for easily finding many solutions.

Key: The digits in the answer must always add up to 18.

Note that this is true for all the examples above. This is a necessary, but not sufficient condition. 459 (173 + 286) is the smallest number for which this works. There are only three other cases (684, 756, and 765) above 459 that will not work. This is a high interest activity that can be presented in a variety of ways.

Additional Activity: These 9 digit cards may also be used to make a 3 X 3 "Magic Square" in which each of the rows, columns, and diagonals adds up to 15. A solution is shown:

8	1	6
3	5	7
4	9	2

Crossword Puzzle

This crossword puzzle is an effective way to review the material covered in this volume of *Historical Connections*. It may also be used to stimulate questions and interest in the lives and accomplishments of the mathematicians mentioned.

Solution:

Magic Card Game

Duplicate the six cards on light card stock and allow students to cut them out so each has a set of his/her own. Point out that this game is based on the binary (base 2) number system. Base 2 has applications in various branches of mathematics, including computers. Some teachers may wish to take students through the process of making the magic cards before giving them the completed game. The activity "Making Magic Cards" is included for this purpose.

"Mathematicians" Game

This game, played like the familiar "Authors," provides students with an enjoyable way to become acquainted with the names and accomplishments of the great mathematicians. Each of the mathematicians in the game has been featured in either Vol. I or Vol. II of *Historical Connections*.

The cards should be photocopied onto light card stock. Some teachers may wish to laminate these before cutting out the cards. Instructions for playing the game are provided on the instruction card.

Note: These cards may also be used to play "Who Am I?" Students may be offered one clue at a time to see how quickly they can identify the mathematician being described. This may be played in pairs or in teams.

RESOURCES FOR LIBRARY AND CLASSROOM

Abbott, David
THE BIOGRAPHICAL DICTIONARY OF SCIENTISTS: MATHEMATICIANS
New York: Peter Bedrick Books, 1986
   This is an authoritative and accessible reference work. Includes a chronological introduction and alphabetical arrangement of entries, plus a useful glossary.

Alic, Margaret
HYPATIA'S HERITAGE
London: The Women's Press, 1990.
   Biographical and scientific information on the lives and times of outstanding women scientists and mathematicians from antiquity to the late 19th century.

Beckman, Petr
A HISTORY OF PI
New York: St. Martin's Press, 1971
   Readable, interesting source which reveals the background of the times and the personalities associated with the development of pi.

Bell, E.T.
MEN OF MATHEMATICS
New York: Simon & Schuster, 1965
   The classic work in mathematics history. Includes lots of detail and useful information.

Dunham, William
JOURNEY THROUGH GENIUS: THE GREAT THEOREMS OF MATHEMATICS
New York: John Wiley & Sons, Inc., 1990
   This work explores some of the most significant and enduring ideas in mathematics, emphasizing the humanity of the great mathematicians.

Eves, Howard W.
AN INTRODUCTION TO THE HISTORY OF MATHEMATICS
Fifth Edition
New York: Saunders College Publishing, 1983
   A popular text for history of math classes. Eves traces the development of mathematics with good humor and informative detail.

Eves, Howard W.
IN MATHEMATICAL CIRCLES (VOL 1 & 2)
Boston: Prindle, Weber & Schmidt, Inc., 1969
   These popular books contain chronologically arranged anecdotes about mathematicians and their discoveries. Delightful, short bits of information which are useful and understandable.

Grinstein, Louise and Paul Campbell
WOMEN OF MATHEMATICS: A BIBLIOGRAPHIC SOURCEBOOK
New York: Greenwood Press, 1987.
   Contains brief biographies of 43 women mathematicians, plus comments on their work and useful bibliographic sources.

Infeld, Leopold
WHOM THE GODS LOVE: THE STORY OF EVARISTE GALOIS
Reston, Virginia: NCTM, 1978
   A fictional biography that explores the life and mathematical pursuits of Galois. Readable, valuable resource.

Hollingdale, Stuart
MAKERS OF MATHEMATICS
London: Penguin Books, 1989
   Chapters on mathematicians from Pythagoras to Einstein provide useful biographical information, accompanied by a solid review of the mathematics these persons worked with.

Kanigel, Robert
THE MAN WHO KNEW INFINITY: A LIFE OF THE GENIUS RAMANUJAN
New York: Charles Scribner's Sons, 1991.
   A biography of the brilliant, self-taught Indian mathematician whose work contains some of the most remarkable ideas in the history of science.

Katz, Victor
A HISTORY OF MATHEMATICS
New York, New York: HarperCollins, 1993
   A readable college text, filled with useful information and illustrations.

Mitchell, Merle
MATHEMATICAL HISTORY: ACTIVITIES, PUZZLES, STORIES, AND GAMES
Reston, Virginia: National Council of Teachers of Mathematics, 1978
   This is a collection of enrichment resources for use in the upper elementary grades. Activities may be photocopied for classroom use.

MULTICULTURALISM IN MATHEMATICS, SCIENCE, AND TECHNOLOGY: READINGS AND ACTIVITIES
Menlo Park, CA: Addison-Wesley, 1993
   A source book for teachers, including representation from a wide range of cultural groups. Black line masters may be reproduced for classroom use.

National Council of Teachers of Mathematics
HISTORICAL TOPICS FOR THE MATHEMATICS CLASSROOM
Reston, Virginia: NCTM, 1989
   This work, first commissioned in 1969, is designed to help teachers teach mathematics from a historical perspective. It is divided into chapters on the history of numbers, the history of geometry, the history of algebra, etc., and includes a useful list of resources.

Pappas, Theoni
THE JOY OF MATHEMATICS
San Carlos, CA: Wide World Publishing/Tetra, 1989
   This book unveils the inseparable relationship of mathematics to the world in which we live. In one or two page "glimpses," the reader enjoys games, puzzles, interesting facts, and historic background.

Pappas, Theoni
MORE JOY OF MATHEMATICS
San Carlos, CA: Wide World Publishing/Tetra, 1991
   Like Pappas' first book, this collection also provides brief but fascinating information on how mathematics can be seen in nature, science, music, architecture, literature, and history.

Perl, Teri
MATH EQUALS—BIOGRAPHIES OF WOMEN MATHEMATICIANS
Menlo Park, CA: Addison-Wesley, 1978
   This is a readable collection of resources on the lives and work of nine women, including activities that relate to their work.

Reimer, Luetta and Wilbert Reimer
MATHEMATICIANS ARE PEOPLE, TOO, Volumes One and Two
Palo Alto, CA: Dale Seymour Publications, 1990, 1995
   This collection of illustrated stories dramatically recreate episodes from the lives of 30 mathematicians, including seven women. For students to read or for teachers to read aloud, the books highlight the human element in mathematics. Appropriate for students in grade three through secondary school.

Reimer, Wilbert and Luetta Reimer
HISTORICAL CONNECTIONS IN MATHEMATICS, Volume I
Fresno, CA: AIMS Education Foundation, 1992, 2005
   Biographical information about ten great mathematicians, plus portraits, illustrations, classroom-ready activities, and suggestions for use. May be reproduced. Complete solutions and a source list. For grades 4 - 10.

Schaaf, William
MATHEMATICS AND SCIENCE: AN ADVENTURE IN POSTAGE STAMPS
Reston, Virginia: National Council of Teachers of Mathematics, 1978
   This book traces, through illustration and historical insight, the way postage stamps mirror the impact of mathematics and science on society.

The AIMS Program

AIMS is the acronym for "**A**ctivities **I**ntegrating **M**athematics and **S**cience." Such integration enriches learning and makes it meaningful and holistic. AIMS began as a project of Fresno Pacific University to integrate the study of mathematics and science in grades K-9, but has since expanded to include language arts, social studies, and other disciplines.

AIMS is a continuing program of the non-profit AIMS Education Foundation. It had its inception in a National Science Foundation funded program whose purpose was to explore the effectiveness of integrating mathematics and science. The project directors, in cooperation with 80 elementary classroom teachers, devoted two years to a thorough field-testing of the results and implications of integration.

The approach met with such positive results that the decision was made to launch a program to create instructional materials incorporating this concept. Despite the fact that thoughtful educators have long recommended an integrative approach, very little appropriate material was available in 1981 when the project began. A series of writing projects ensued, and today the AIMS Education Foundation is committed to continuing the creation of new integrated activities on a permanent basis.

The AIMS program is funded through the sale of books, products, and professional-development workshops, and through proceeds from the Foundation's endowment. All net income from programs and products flows into a trust fund administered by the AIMS Education Foundation. Use of these funds is restricted to support of research, development, and publication of new materials. Writers donate all their rights to the Foundation to support its ongoing program. No royalties are paid to the writers.

The rationale for integration lies in the fact that science, mathematics, language arts, social studies, etc., are integrally interwoven in the real world, from which it follows that they should be similarly treated in the classroom where students are being prepared to live in that world. Teachers who use the AIMS program give enthusiastic endorsement to the effectiveness of this approach.

Science encompasses the art of questioning, investigating, hypothesizing, discovering, and communicating. Mathematics is a language that provides clarity, objectivity, and understanding. The language arts provide us with powerful tools of communication. Many of the major contemporary societal issues stem from advancements in science and must be studied in the context of the social sciences. Therefore, it is timely that all of us take seriously a more holistic method of educating our students. This goal motivates all who are associated with the AIMS Program. We invite you to join us in this effort.

Meaningful integration of knowledge is a major recommendation coming from the nation's professional science and mathematics associations. The American Association for the Advancement of Science in *Science for All Americans* strongly recommends the integration of mathematics, science, and technology. The National Council of Teachers of Mathematics places strong emphasis on applications of mathematics found in science investigations. AIMS is fully aligned with these recommendations.

Extensive field testing of AIMS investigations confirms these beneficial results:

1. Mathematics becomes more meaningful, hence more useful, when it is applied to situations that interest students.
2. The extent to which science is studied and understood is increased when mathematics and science are integrated.
3. There is improved quality of learning and retention, supporting the thesis that learning which is meaningful and relevant is more effective.
4. Motivation and involvement are increased dramatically as students investigate real-world situations and participate actively in the process.

We invite you to become part of this classroom teacher movement by using an integrated approach to learning and sharing any suggestions you may have. The AIMS Program welcomes you!

AIMS Education Foundation Programs

When you host an AIMS workshop for elementary and middle school educators, you will know your teachers are receiving effective, usable training they can apply in their classrooms immediately.

AIMS Workshops are Designed for Teachers
- Correlated to your state standards;
- Address key topic areas, including math content, science content, and process skills;
- Provide practice of activity-based teaching;
- Address classroom management issues and higher-order thinking skills;
- Give you AIMS resources; and
- Offer optional college (graduate-level) credits for many courses.

AIMS Workshops Fit District/Administrative Needs
- Flexible scheduling and grade-span options;
- Customized (one-, two-, or three-day) workshops meet specific schedule, topic, state standards, and grade-span needs;
- Prepackaged four-day workshops for in-depth math and science training available (includes all materials and expenses);
- Sustained staff development is available for which workshops can be scheduled throughout the school year;
- Eligible for funding under the Title I and Title II sections of No Child Left Behind; and
- Affordable professional development—consecutive-day workshops offer considerable savings.

University Credit—Correspondence Courses
AIMS offers correspondence courses through a partnership with Fresno Pacific University.
- Convenient distance-learning courses—you study at your own pace and schedule. No computer or Internet access required!

Introducing AIMS State-Specific Science Curriculum
Developed to meet 100% of your state's standards, AIMS' State-Specific Science Curriculum gives students the opportunity to build content knowledge, thinking skills, and fundamental science processes.
- Each grade-specific module has been developed to extend the AIMS approach to full-year science programs. Modules can be used as a complete curriculum or as a supplement to existing materials.
- Each standards-based module includes math, reading, hands-on investigations, and assessments.

Like all AIMS resources, these modules are able to serve students at all stages of readiness, making these a great value across the grades served in your school.

For current information regarding the programs described above, please complete the following form and mail it to: P.O. Box 8120, Fresno, CA 93747.

Information Request

Please send current information on the items checked:

___ *Basic Information Packet* on AIMS materials
___ Hosting information for AIMS workshops
___ AIMS State-Specific Science Curriculum

Name: _____

Phone: _____ E-mail: _____

Address: _____
Street   City   State   Zip

AIMS Magazine

YOUR K-9 MATH AND SCIENCE CLASSROOM ACTIVITIES RESOURCE

The AIMS Magazine is your source for standards-based, hands-on math and science investigations. Each issue is filled with teacher-friendly, ready-to-use activities that engage students in meaningful learning.

- *Four issues each year (fall, winter, spring, and summer).*

Current issue is shipped with all past issues within that volume.

1825	Volume XXV	2010-2011	$19.95
1826	Volume XXVI	2011-2012	$19.95

Two-Volume Combinations
M20911	Volumes XXIV & XXV	2009-2011	$34.95
M21012	Volumes XXV & XXVI	2010-2012	$34.95

Complete volumes available for purchase:

1802	Volume II	1987-1988	$19.95
1804	Volume IV	1989-1990	$19.95
1805	Volume V	1990-1991	$19.95
1807	Volume VII	1992-1993	$19.95
1808	Volume VIII	1993-1994	$19.95
1809	Volume IX	1994-1995	$19.95
1810	Volume X	1995-1996	$19.95
1811	Volume XI	1996-1997	$19.95
1812	Volume XII	1997-1998	$19.95
1813	Volume XIII	1998-1999	$19.95
1814	Volume XIV	1999-2000	$19.95
1815	Volume XV	2000-2001	$19.95
1816	Volume XVI	2001-2002	$19.95
1817	Volume XVII	2002-2003	$19.95
1818	Volume XVIII	2003-2004	$19.95
1819	Volume XIX	2004-2005	$19.95
1820	Volume XX	2005-2006	$19.95
1821	Volume XXI	2006-2007	$19.95
1822	Volume XXII	2007-2008	$19.95
1823	Volume XXIII	2008-2009	$19.95
1824	Volume XXIV	2009-2010	$19.95

Volumes II to XIX include 10 issues.

Call 1.888.733.2467 or go to www.aimsedu.org

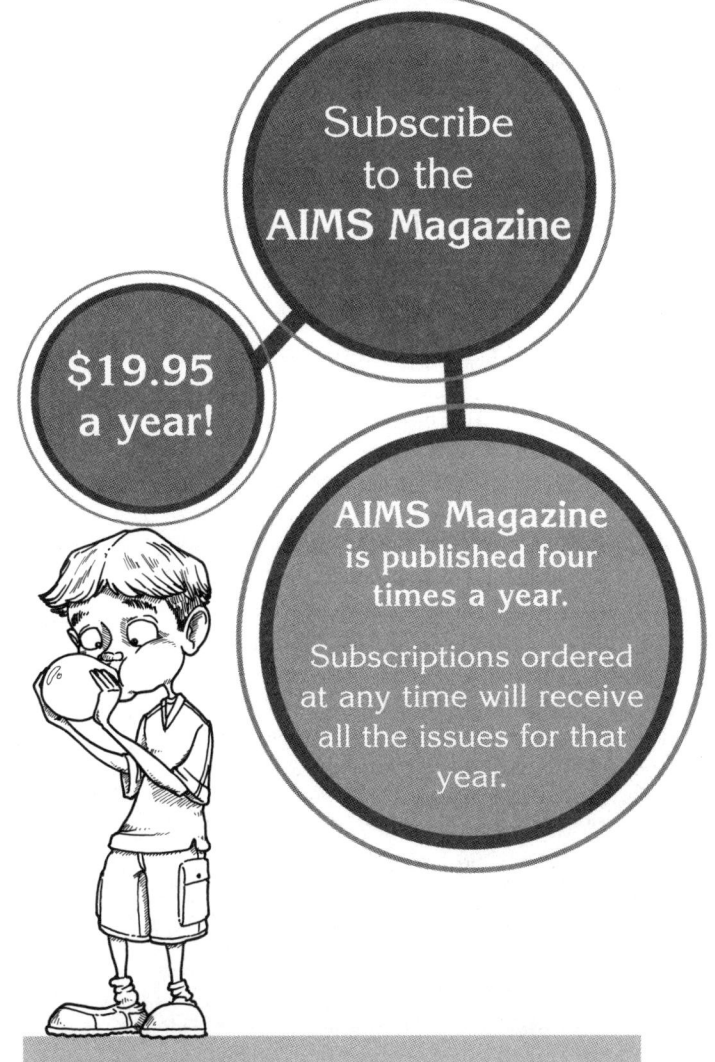

Subscribe to the AIMS Magazine

$19.95 a year!

AIMS Magazine is published four times a year.

Subscriptions ordered at any time will receive all the issues for that year.

AIMS Online—www.aimsedu.org

To see all that AIMS has to offer, check us out on the Internet at www.aimsedu.org. At our website you can preview and purchase AIMS books and individual activities, learn about State-Specific Science and Essential Math, explore professional development workshops and online learning opportunities, search our activities database, buy manipulatives and other classroom resources, and download free resources including articles, puzzles, and sample AIMS activities.

AIMS E-mail Specials
While visiting the AIMS website, sign up for our FREE e-mail newsletter with subscriber-only specials. You'll also receive advance notice of new products.

Sign up today!

AIMS Program Publications

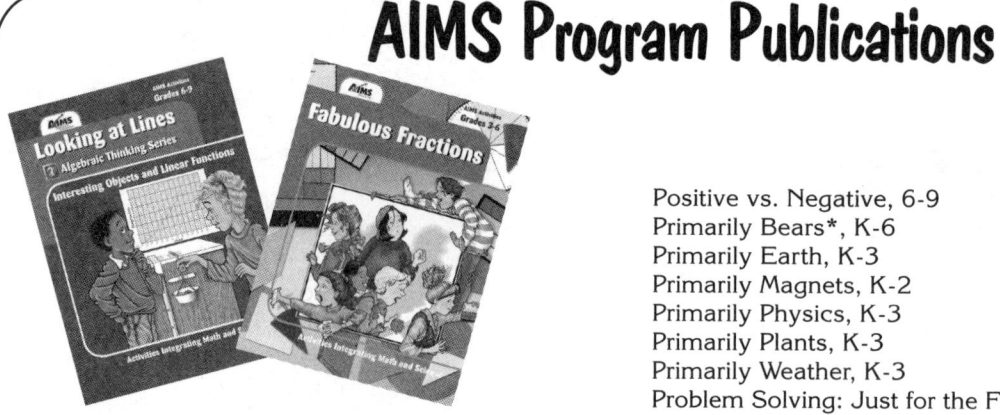

Actions With Fractions, 4-9
The Amazing Circle, 4-9
Awesome Addition and Super Subtraction, 2-3
Bats Incredible! 2-4
Brick Layers II, 4-9
The Budding Botanist, 3-6
Chemistry Matters, 5-7
Counting on Coins, K-2
Cycles of Knowing and Growing, 1-3
Crazy About Cotton, 3-7
Critters, 2-5
Earth Book, 6-9
Electrical Connections, 4-9
Exploring Environments, K-6
Fabulous Fractions, 3-6
Fall Into Math and Science*, K-1
Field Detectives, 3-6
Finding Your Bearings, 4-9
Floaters and Sinkers, 5-9
From Head to Toe, 5-9
Glide Into Winter With Math and Science*, K-1
Gravity Rules! 5-12
Hardhatting in a Geo-World, 3-5
Historical Connections in Mathematics, Vol. I, 5-9
Historical Connections in Mathematics, Vol. II, 5-9
Historical Connections in Mathematics, Vol. III, 5-9
It's About Time, K-2
It Must Be A Bird, Pre-K-2
Jaw Breakers and Heart Thumpers, 3-5
Looking at Geometry, 6-9
Looking at Lines, 6-9
Machine Shop, 5-9
Magnificent Microworld Adventures, 6-9
Marvelous Multiplication and Dazzling Division, 4-5
Math + Science, A Solution, 5-9
Mathematicians are People, Too
Mathematicians are People, Too, Vol. II
Mostly Magnets, 3-6
Movie Math Mania, 6-9
Multiplication the Algebra Way, 6-8
Out of This World, 4-8
Paper Square Geometry:
    The Mathematics of Origami, 5-12
Puzzle Play, 4-8
Popping With Power, 3-5

Positive vs. Negative, 6-9
Primarily Bears*, K-6
Primarily Earth, K-3
Primarily Magnets, K-2
Primarily Physics, K-3
Primarily Plants, K-3
Primarily Weather, K-3
Problem Solving: Just for the Fun of It! 4-9
Problem Solving: Just for the Fun of It! Book Two, 4-9
Proportional Reasoning, 6-9
Ray's Reflections, 4-8
Sensational Springtime, K-2
Sense-able Science, K-1
Shapes, Solids, and More: Concepts in Geometry, 2-3
The Sky's the Limit, 5-9
Soap Films and Bubbles, 4-9
Solve It! K-1: Problem-Solving Strategies, K-1
Solve It! 2nd: Problem-Solving Strategies, 2
Solve It! 3rd: Problem-Solving Strategies, 3
Solve It! 4th: Problem-Solving Strategies, 4
Solve It! 5th: Problem-Solving Strategies, 5
Solving Equations: A Conceptual Approach, 6-9
Spatial Visualization, 4-9
Spills and Ripples, 5-12
Spring Into Math and Science*, K-1
Statistics and Probability, 6-9
Through the Eyes of the Explorers, 5-9
Under Construction, K-2
Water, Precious Water, 4-6
Weather Sense: Temperature, Air Pressure, and Wind, 4-5
Weather Sense: Moisture, 4-5
What's Next, Volume 1, 4-12
What's Next, Volume 2, 4-12
What's Next, Volume 3, 4-12
Winter Wonders, K-2

Essential Math
Area Formulas for Parallelograms, Triangles, and Trapezoids, 6-8
Circumference and Area of Circles, 5-7
Effects of Changing Lengths, 6-8
Measurement of Prisms, Pyramids, Cylinders, and Cones, 6-8
Measurement of Rectangular Solids, 5-7
Perimeter and Area of Rectangles, 4-6
The Pythagorean Relationship, 6-8

Spanish Edition
Constructores II: Ingeniería Creativa Con Construcciones
    LEGO®, 4-9
    The entire book is written in Spanish. English pages not included.

* Spanish supplements are available for these books. They are only available as downloads from the AIMS website. The supplements contain only the student pages in Spanish; you will need the English version of the book for the teacher's text.

For further information, contact:
AIMS Education Foundation • P.O. Box 8120 • Fresno, California 93747-8120
www.aimsedu.org • 559.255.6396 (fax) • 888.733.2467 (toll free)

Duplication Rights

No part of any AIMS books, magazines, activities, or content—digital or otherwise—may be reproduced or transmitted in any form or by any means—including photocopying, taping, or information storage/retrieval systems—except as noted below.

Standard Duplication Rights

- A person or school purchasing AIMS activities (in books, magazines, or in digital form) is hereby granted permission to make up to 200 copies of any portion of those activities, provided these copies will be used for educational purposes and only at one school site.
- Workshop or conference presenters may make one copy of any portion of a purchased activity for each participant, with a limit of five activities per workshop or conference session.
- All copies must bear the AIMS Education Foundation copyright information.

Standard duplication rights apply to activities received at workshops, free sample activities provided by AIMS, and activities received by conference participants.

Unlimited Duplication Rights

Unlimited duplication rights may be purchased in cases where AIMS users wish to:
- make more than 200 copies of a book/magazine/activity,
- use a book/magazine/activity at more than one school site, or
- make an activity available on the Internet (see below).

These rights permit unlimited duplication of purchased books, magazines, and/or activities (including revisions) for use at a given school site.

Activities received at workshops are eligible for upgrade from standard to unlimited duplication rights.

Free sample activities and activities received as a conference participant are not eligible for upgrade from standard to unlimited duplication rights.

State-Specific Science modules are licensed to one classroom/one teacher and are therefore not eligible for upgrade from standard to unlimited duplication rights.

Upgrade Fees

The fees for upgrading from standard to unlimited duplication rights are:
- $5 per activity per site,
- $25 per book per site, and
- $10 per magazine issue per site.

The cost of upgrading is shown in the following examples:
- activity: 5 activities x 5 sites x $5 = $125
- book: 10 books x 5 sites x $25 = $1250
- magazine issue: 1 issue x 5 sites x $10 = $50

Purchasing Unlimited Duplication Rights

To purchase unlimited duplication rights, please provide us the following:
1. The name of the individual responsible for coordinating the purchase of duplication rights.
2. The title of each book, activity, and magazine issue to be covered.
3. The number of school sites and name of each site for which rights are being purchased.
4. Payment (check, purchase order, credit card)

Requested duplication rights are automatically authorized with payment. The individual responsible for coordinating the purchase of duplication rights will be sent a certificate verifying the purchase.

Internet Use

AIMS materials may be made available on the Internet if all of the following stipulations are met:
1. The materials to be put online are purchased as PDF files from AIMS (i.e., no scanned copies).
2. Unlimited duplication rights are purchased for all materials to be put online for each school at which they will be used. (See above.)
3. The materials are made available via a secure, password-protected system that can only be accessed by employees at schools for which duplication rights have been purchased.

AIMS materials may not be made available on any publicly accessible Internet site.

HISTORICAL CONNECTIONS VOL. II © 2005 AIMS Education Foundation